Farouk Benaceur
Kenza Benaceur

Microbial genetics mini-manual

Farouk Benaceur
Kenza Benaceur

Microbial genetics mini-manual

Theoretical foundations & corrected exercises

ScienciaScripts

Imprint

Cover image: www.ingimage.com

This book is a translation from the original published under ISBN 978-620-6-72249-6.

Publisher:
Sciencia Scripts
is a trademark of
Dodo Books Indian Ocean Ltd. and OmniScriptum S.R.L publishing group

120 High Road, East Finchley, London, N2 9ED, United Kingdom
Str. Armeneasca 28/1, office 1, Chisinau MD-2012, Republic of Moldova, Europe
Printed at: see last page
ISBN: 978-620-8-20644-4

Contents

Foreword

Microbial gënëtics is a very important field involving the study of the relationship between the foundations of gënëtics and those of microbiology, as well as an in-depth understanding of how the molecular machinery of microorganisms works.

This handout constitutes a practical rëfëference in microbial gënëtics intended for ëtudents interested in the study of microbiology ttpplied to microbial biotechnology, molecular biology, and genetics.

The handout is organised into four well-detailed chapters covering all the basic concepts of microbial genetics, including an introductory chapter on genetic material and replication, a second chapter on the expression of genetic information and its regulation, a third on mutations and repair systems, and a fourth on gene exchange and transfer.

In addition, there is a collection of a considerable number of varied and targeted exercises including reflection and calculation questions and a description of practical situations often encountered in the laboratory.

Nine series of tutorials including 55 well-defined exercises with solutions to more than 35 exercises; the difference in structural, metabolic and genetic exchanges between prokaryotes and eukaryotes, with the complexity of the phenomënes involved in vital processes involving the continuity, survival and/or transmission of genetic material attracting increasing attention in order to understand both the basis of genome genënes and the degree of ingeniosity involved.

The study of mutations, inutagenesis and repair systems will also be covered. This provides a concrete and realistic idea of the intrinseal and extrinseal disturbances that can modify a genotype or phenotype, giving rise to negative or positive consequences. It also clarifies a number of confusing issues with scientific logic and practical proof.

Genetic exchange, including recombination, conjugation, and generalized and specialized transduction, are also very interesting phenomena that describe the modes of transfer, transmission and acquisition of new characteristics, and open up a wide range of applications.

In this mimeo, a brief reminder of the necessary prerequisites and the objectives will be presented before each series of exercises. It constitutes a practical mini-manual ensuring a good understanding of the course concepts via explanations and methodical correction of exercises and problems. It is intended for students of several undergraduate disciplines including Applied Microbiology, Microbial Biotechnology, Molecular Biology, and Genetics.

Introduction

The study of the diversity of the microbial world is attracting more and more attention from microbiologists and geneticists, who are keen to reveal the secrets of evolution, and to uncover the secrets of these organisms, which are tiny in terms of size but complicated by their structural organisation and metabolism.

The evolutionary study of micro-organisms is called phylogeny. Understanding the phylogenetic profiles of microorganisms is a difficult task because of their small size and the lack of specific indicators that can be used as benchmarks. Certain proteins, genes and consensus sequences are considered to be evolutionary indices that describe the state of evolutionary change. Currently, analysis of the 16S rDNA sequence is considered the most reliable method for measuring evolutionary relationships between bacteria and archaea, while the 18S sequence is used in the case of microscopic fungi.

At present, the classification of bacteria is based on taking into account as many data as possible: genetic data and phylogenetic constructs, but also and always phenotypic data and ecological data, hence the term "Polyphasic Taxonomy" or even "mixed and consensual taxonomy".

The phylogenetic tree representing all living organisms shows that the evolution of current forms of life took place from a common ancestor (the universal ancestor), represented by the root (Figure 1). Two areas make up the prokaryotic life systems: the archaea and the eubacteria.

However, it is interesting to note that genomic studies have shown that archaea possess unique gene sequences that are not present in bacteria or eukaryotes. Certain genes are also shared between the three domains. The genes required for the cellular functions of the nucleus are those that are necessary for the survival of a cell and that could have come from the common ancestor.

The divergence of organisms represents the differences in gene sequences retained in each group as they evolved. It is also postulated that HGT (horizontal gene transfer) played a key role in the transfer of genes between organisms at the beginning of evolutionary history.

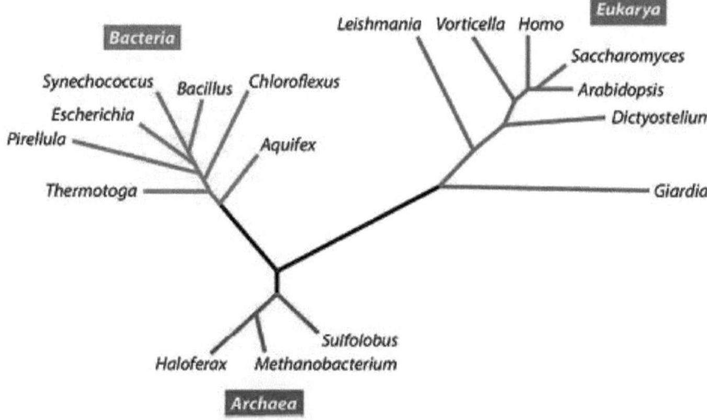

Synechococcus Bacillus Chloroflexus
Escherichia
Pirellula
Thermotoga

\ *Sulfolobus*
Methanobacterium
Figure 1: Phylogenetic tree of Microorgnisms based on SSU rRNA sequences.

2-Structural organisation

The various groups of living organisms must be studied in relation to their structural unit, the cell. In 1937, the protozoologist Edouard Chatton proposed the terms "eukaryote" and "prokaryote", but the true nature of these two cell types was not clearly defined until 1955, with the progress of electron microscopy.

A distinction is made between the eukaryotic cell, characteristic of plants, animals, higher protists and microscopic fungi, and the prokaryotic cell, characteristic of lower protists, particularly bacteria.

Several differences exist between these two types of cell, taking into account several comparison criteria. A brief summary of these differences is shown in Table 1.

In addition, the discipline of microbial genetics is much more interested in the analysis of genetic material from the point of view of structure, organisation and the exchanges that can take place.

The main genetic material of eukaryotic organisms is made up of chromosomes whose basic element is, as in prokaryotic organisms, a double helix of DNA. The chromosomes are enclosed in a defined cellular structure called the nucleus.

The nucleus is surrounded by a nuclear membrane which contains pores and is linked to the cell's system of internal membranes. The nucleus contains an RNA-rich structure called a nucleolus: this structure is only visible at certain stages of the cell cycle and is difficult or impossible to observe in many eukaryotic microorganisms (it can be seen in protozoa).

Table 1: Comparison between prokaryotic and eukaryotic cells

Properties	Prokaryotes	Eukaryotes
Organisation of genetic material		
-Core bounded by a veritable	Absent	Present
Histone-bound DNA envelope	No	Yes
-Number of chromosomes	One (genophore)	More than one
-Introns	Rare	Brakes
- Nucleole	Absent	Absent
-Mitosis	No	Yes
-Histones	LHP (Histone-like proteins)	Present
Genetic recombination	Partial and unidirectional DNA transfer	Meiosis and fusion
Mitochondria	Absent	Presentation
Chloroplasts	Absent	Presentation
Plasma membrane with sterols	Not usually	Yes
Flagella	Submicroscopic in size; composed of a single fibre	The microscopic size of the membrane
Endoplasmic reticulum	Absent	Present
Golgi apparatus	Absent	Present
Cell wall	Chemically complex with doglycan peptides	Chemically simple with no peptidogly- canes
Ribosomes	Free 70S	Bound and free 80S

On the other hand, microbial cells store most of their genetic information in double-stranded DNA. It should be noted, however, that information can sometimes be carried by single-stranded DNA in extrachromosomal structures ("parasitic" viruses, for example).

Duplex DNA is formed by the assembly of two polynucleotide chains (or strands) of DNA

joined by hydrogen bonds that unite the bases. Base pairing is well defined: adënine - thymine (A-T) and cytosine - guanine (C-G).
The bond between G and C uses 3 hydrogen bridges, while that between A and T uses two: naturally, the G-C bond is stronger and G-C-dominant double bonds are more stable than others.
There is also extrachormosomal gëtic material, including organelle DNA (mitochondrial DNA and chlorplast DNA), plasmids (bacterial and eukaryotic), yeast killer factors and paramëcies, and bacteriophage.
In the following section, we will look at this structural difference in more detail, with an exclusive focus on genetic material. Several key and specialised words will be quoted, and we present below a sort of glossary containing the definitions of some of the terms used in microbial genetics.
Gene: unit of hereditary transmission of genetic information. A gëne is a segment of DNA (or RNA in viruses), located at a precise locus on a chromosome, which includes the sequence coding for a protein, and the sequences which allow and regulate its expression.
Genome: all the genetic material present in each of an individual's cells. An individual's hereditary heritage.
Genotype: all the genetic characteristics of an individual. Its expression leads to the phenotype.
Intron: fragment of a gene located between two exons. Introns are present in immature mRNA and absent in mature mRNA. Non-coding" fragment of the gene.
Locus: precise location of a particular gene on a chromosome.
Mutagen: agent (physical) or substance (chemical) capable of producing mutations. It increases the frequency of certain mutations.
Directed mutagenesis: introduction of a precise mutation into a clone DNA fragment, followed by reinsertion of the mutated sequence into the original gene to replace the corresponding wild-type DNA.
Insertional mutagenesis: introduction of mutations into a cloned or uncloned DNA sequence by insertion of a foreign DNA fragment. This introduction is most often carried out in vivo by insertion of a transposable element which allows the repëraдe of the gene thus mutated.
Mutant: virus, cell or individual possessing a gene that has undergone mutation.
Temperate phage: a phage whose genome can integrate into the host cell's DNA and transform its properties.1 Integrated into the host cell's genome, the phage is called a prophage. 2. The temperate phage is capable of lysogenising the bacteria it infects.
Transducing phage: phage capable of transmitting part of the genome from one host cell to another.
Virulent phage: phage that produces an infection in the bacterium leading to the lytic cycle. The bacteria are lysed and the newly synthesised viral particles are released.
Phenotype: all the observable characteristics of an individual, resulting from the interaction between its genotype and the effects of its environment.
Lysis plate: hole formed in a layer of bacteria following a lytic infection caused by a bacteriophage. The characteristics of the patches (rate of formation, size, appearance, etc.) are often used to define a given type of phage.
Plasmid: small circular molecule of extrachromosomal DNA present in bacte- ries, capable of autonomous replication in the cell of origin and in a host cell. This molecule carries genetic characteristics that are not essential to the host cell. Some plasmids are used as gene cloning

vectors.

Recombinant plasmid: plasmid into which a fragment of foreign DNA has been inserted. This term is most often used to designate a plasmid created by in vitro recombination.

Promoter: a region of DNA located close to a gene and essential for the transcription of DNA into RNA.

Operon: transcription unit consisting of a promoter, an operator and one or more structural genes.

Repressor: a protein that binds to the regulatory sequences of DNA (or RNA), thereby blocking transcription or translation respectively.

Chapter I

Genetic material and its replication

1-Generalities on genetic material

1.1 Nucleic acids

Nucleic acids are **the basic** building blocks of genetic material. They are made up of linear chains of Hë3 nucleotides joined together **by phosphodiester bridges**.

Nucleotides are formed by a purine or pyrimidine base Hëc to a pentose which is itself linked to a phosphate group (Fig 1).

H O_a

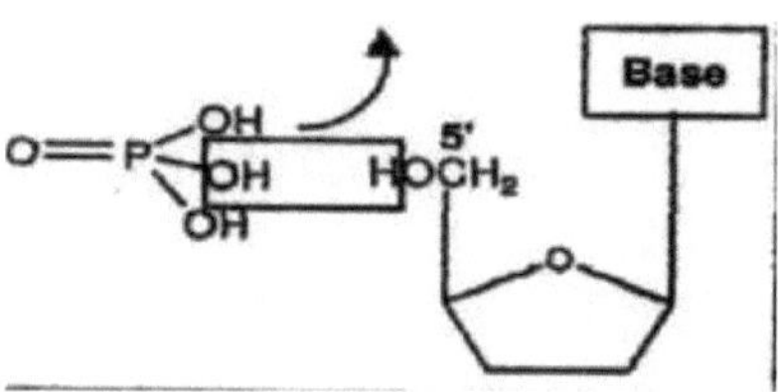

Fig 2 :Formation of the nucleotide by association between a nucleoside (Sugar+ nitrogen base nitrogenous base) and a phosphoric acid.

Phosphoric acid is a tri-acid, with two of its three _or_.,X'ₜ acid functions being esterified in DNA and RNA, OH

Deoxyribonucleic acid (DNA) is formed from **2-desoxyribose** and ribonucleic acid (RNA) from ribose.

Nucleosides result from the association between a nitrogenous base and a pentose via an **N-glycosidic** bond, which is formed between carbon 1 of the ribose and nitrogen 1 of the pyrimidine base or nitrogen 9 of the purine base.

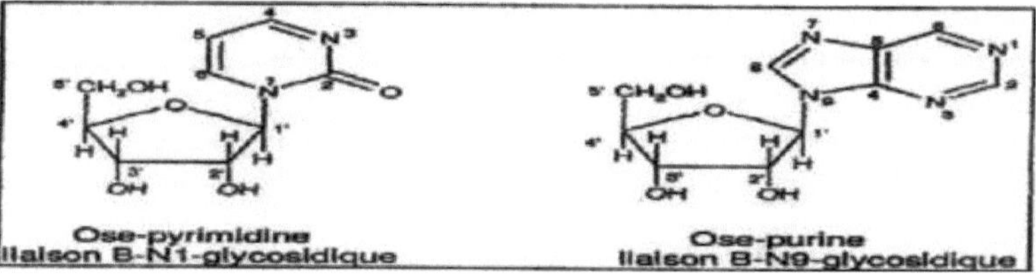

Fig 3: Bonds between pentose and nucleic bases

The names of the nucleosides are those of the root of the base to which is added the desinence "idine" for pyrimidines and "osine" for purines: Cytidinc, Thymidine. Uridine, Adenosinc and Guanosine (Table I)

Table I. Nomenclature of nucleosides and nucleotides making up nucleic acids

		Bases				
		Adenine	Guanine	Cytosine	Uracile	Thymine
Nucleosides	ds RNA	Adenosine	Guanosine	Cytidine	Uridine	-
	dsDNA	d6soxy adenosine	desoxy guanosine	desoxy cytidine	-	ddsoxy thymidine
Nucleotides	ds RNA	Adenylate	Guanylate	Cytidylate	Uridylate	-
	ds DNA	desoxy adenylate	desoxy guanylate	desoxy cytidylate	-	desoxy thymidylate

Nucldoside monophosphate	AMP	GMP	CMP	UMP	-
NucMosides diphosphate	ADP	GDP	CDP	UDP	-
Nucldoside triphosphate	ATP	GTP	CTP	UTP	-
D^sox y nucleoside monophosphate	dAMP	dGMP	dCMP	-	dTMP
Ddsoxynucieoside diphosphate	dADP	dGDP	dCDP	-	dTDP
Ddsoxynucldoside triphosphate	dATP	dGTP	dCTP	-	dTTP

The nitrogen bases of nucleic acids are either pyrimidine or purine. Pyrimidines are aromatic hëtërocycles with six atoms running clockwise from a hëtëroatom (in this case the nitrogen atom).

The purine cycle results from the fusion of two cycles, the pyrimidine cycle and the imidazole cycle.

The bases involved in DNA are **purine bases**: adënine (A) (6- amino purine), and guanine (G) (2- amino, 6- oxy purine), and **pyrimidine bases**: thymine (T)(2, 4-dioxy, 5-mëthyl pyrimidine or 5-mëthyl uracil) and cytosine (C) (2-oxy, 4-amino pyrimidine).... (Fig.). The corresponding nucldosides are adënosine, guanosine, thymidine and cytidine.

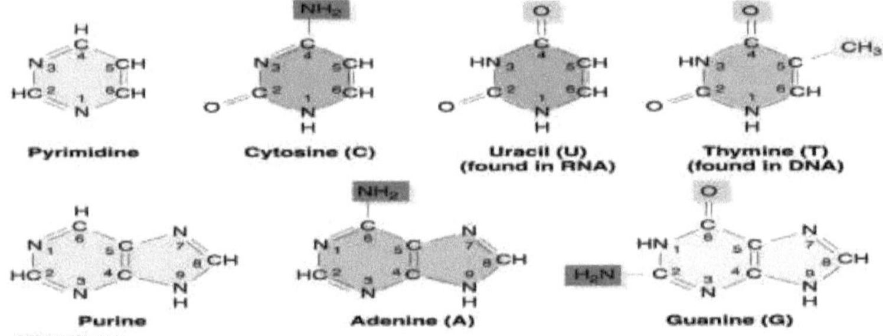

Fig 4:Structure of nitrogenous bases

Certain DNA bases are **modified**: for example, lbydroxymethylcytosine is found in certain bacteriophages, as are significant quantities of 6-methyladenine in prokaryotic organisms and 5-methylcytosine in eukaryotes.

This base **methylation** is found in both prokaryotic and eukaryotic organisms **and is involved in** repairing DNA damage or controlling gene expression (by facilitating transitions in the secondary structure).

RNA contains the same purine bases as DNA, but the pyrimidine bases are represented by uracil (U) (2, 4-dioxy pyrimidine), which replaces thymine, and cytosine. The nucleoside corresponding to uracil is uridine. RNA can also contain modified bases

The cells of protists (as well as those of animals and plants) contain both types of nucleic acid (DNA and RNA), whereas **viruses (bacteriophages) contain** either one or the other (adenovirus and/or retrovirus).

1.2 DNA structure in microbial cells

Microbial cells store most of their genetic information in **double-stranded DNA**. It should be noted, however, that information can sometimes be carried by single-stranded DNA in **extrachromosomal structures** ("parasitic" viruses, for example).

Duplex DNA is formed by the assembly of two DNA "polynucleotide chains" (or strands) joined by hydrogen bonds that unite the bases. Base pairing is 1 well defined: adenine - thymine (A-T) and cytosine - guanine (C-G) The bond between G and C uses 3 hydrogen bridges, that between A and T two: naturally, the G-C bond is stronger and double helices with a predominance of G-C are more stable than others.

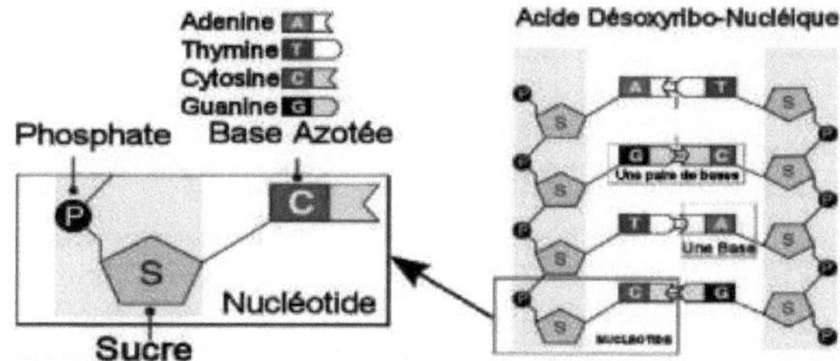

Fig 5 :Simplified DNA structure

The assembly of the chains forms a double hëlice, each strand of which is complementary to the other and antiparallel: one chain is polarized 5'-> 3' and the other 3'-> 5'.

In an aqueous medium, DNA is presented in a stable configuration known as **"B form"**: the direction of winding of this helix is straight and one turn of the helix corresponds to 10 bases. The spacing between two consëcutive bases is **0.34 nm (3.4A°);**The pitch of the helix is 3.4 nm (**34A°**);The diamëtre of 1^ëHcc is 2 nm (20A°);Purine and pyrimidine bases are inside 1^ëHcc (**hydrophobic interior)** and phosphate groups and dësoxyriboses are outside (**hydrophilic exterior).**

DNA molecules can have other types of conformation which are obtained in vitro under certain conditions: **form A** has a right-handed winding of 11 bases per turn and **form Z** has a left-handed winding of 12 bases per turn.

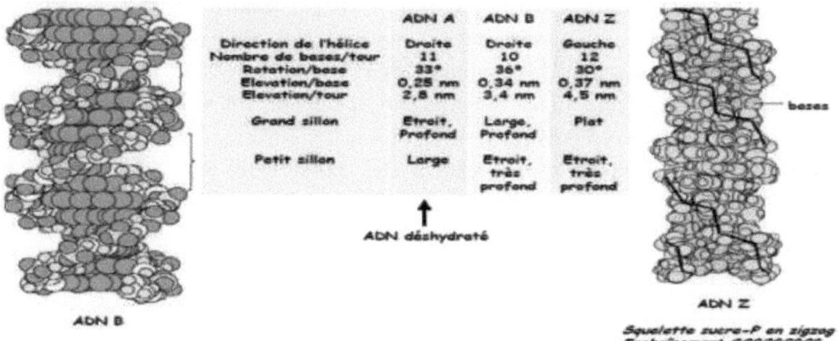

	ADN A	ADN B	ADN Z
Direction de l'hélice	Droite	Droite	Gauche
Nombre de bases/tour	11	10	12
Rotation/base	33°	36°	30°
Elevation/base	0,25 nm	0,34 nm	0,37 nm
Elevation/tour	2,8 nm	3,4 nm	4,5 nm
Grand sillon	Etroit, Profond	Large, Profond	Plat
Petit sillon	Large	Etroit, très profond	Etroit, très profond

Fig 6 :Different forms of DNA

The double helix can be supercoiled either negatively or positively. Supercoiling is said to be negative when it occurs in a direction opposite to the right-hand direction of the double helix: it is very frequent in vivo. Supercoiling is said to be positive when it occurs in the same direction: it only occurs in vitro.

The intermediate, non-supercoiled state is called r el ache or relaxcd. All DNA molecules with the same degree of supercoiling are topological isomers.

The enzymes controlling winding or unwinding at the supercoils are **topoisomerases**, of which there are several types. They act by two types of action: by transient cleavage of one strand (type I) or of two strands (type II)). Type 1 topoisomërases are ATP-dependent. **Gyrase** is a particular type II topoisomërase found in bacteria which introduces negative supercoiling (up to 100/min) in the presence of ATP.

DNA is highly resistant to alkaline hydrolysis. The double helix is easily denatured by heating: the two chains separate and form random loops. Each type of DNA has a characteristic denaturation temperature (or melting point).

The higher the G-C content, the higher the melting point (because of the strong bond between G and C). For a given species, the overall base composition of the DNA is a constant: this composition can be expressed **as the ratio A+T/G+C.**

This ratio varies from one species to another, but is constant and characteristic for a given species: the greatest range of variation is found in bacteria, from 0.25 for *Micrococcus luteus* to 2.9 for *Clostridium perfringens*. "Wallace's rule (Tm calculation for small fragments): Tm = 4*(number of GC bases) + 2*(number of AT bases) Ex: AGTCATGCCCCGCGC Tm = 4*11 +2*4 = 52°C "

The DNA content varies greatly from one organism to another: the contents are 0.2, 4, 50 and 31.10^{-15} g respectively for bacteriophage T4 (virus), the bacterium *Escherichia coli* (prokaryotic organism), the yeast *Saccharomyces cerevisiae* and the protozoan *Euglena gracilis* (eukaryotic organisms). The number of nucleotide pairs varies from 10^3 to 10^5 for viruses (1 to 200 genes), from 10^3 to 10^{+5} for bacteria (1,200 to 3,000 genes) and from 10^{+7} to 10^{+9} for eukaryotic micro-organisms (3.10^3 to 10.10^3 genes).

The kilobase (kb) is usually used to measure the number of nucleotides. For double-stranded nucleic acids, a kilobase corresponds to **103** base pairs (bp) or, to put it another way, nucleotides. We consider that **1 kb** corresponds to a molecular mass **of 0.7.106 Da** and a size of **0.35 gm.**

1.3 Organisation of genetic material at cell level

1.3.1 Genetic material of prokaryotic organisms: the genophore

In prokaryotic organisms, the main genetic material is a single **double-stranded DNA structure** called a chromosome or genophore: this structure is located in an area of the cell called the nucleus or nucleoid. In addition to DNA, it contains RNA (up to 30%) and proteins (10%), most of which is RNA polymërase.

The chromosome is **circular,** i.e. it is closed. The double helix is wound into a superhelix (400 base pairs per turn) which is itself folded to form loops held together by a support consisting mainly of RNA.

In *Escherichia coli*, there are around fifty loops corresponding to 200 turns of the superhelix. The deployed chromosome is up to 1 mm long, so it is extremely tightly packed in the cell...

DNA is complexed with Mg^{2+} or Ca^{2+} cations and with proteins. Some (protein P) are of the protamine type (similar to spermine or spermidine), which neutralise the negative charges on the DNA. In *Escherichia coli*, basic and thermostable proteins similar to the histones of eukaryotic organisms (like histon protein) have been reported. There are at least four of these: Hu (formed from 2 o and в subunits close to histone H2B), H (close to histone H2A), H1 and HLP1 (which are thought to play a role in the non-specific regulation of transcription)...

Morphological and biochemical studies show that DNA is bound to the cytoplasmic membrane or to invaginations of it called mesosomes. In prokaryotic organisms, the chromosome is made up of a continuous chain of gënes and does not contain large repetitive regions (except for RNA gënes).

1.3.2 Genetic material of eukaryotic organisms: chromosomes 1.3.2.1 Generalities

The main genetic material of eukaryotic organisms is made up of chromosomes whose basic element is, as in prokaryotic organisms, a double helix of DNA. The chromosomes are enclosed in a defined cellular structure called the nucleus, **which** is surrounded by a nuclear membrane containing pores and is linked to the cell's internal membrane system. The nucleus contains an RNA-rich structure called the **nucleolus**: this structure is only visible at certain stages of the cell cycle and is difficult or impossible to observe in many eukaryotic microorganisms (it can be seen in protozoa).

Another tubule-shaped structure is sometimes visible in the cytoplasm near the nucleus or is integral with the nuclear membrane: it corresponds to the classical **centrosome** of the nucleus of animal cells or the diffuse centrosome of plant cells. In many eukaryotic microorganisms, this structure is present in a more or less clear manière, and is usually appelëe spindle plate. This structure is sometimes only observable by ëlectron microscopy. .

The number of nuclei is, in principle, one per cell, but there are many exceptions in microorganisms, and cells with several nuclei are often found, independently of the transitory situations resulting from sexual^. In Paramecium (a ciliated protozoan), there are **two types** of nucleus (the macro- and the micronucleus) with different functions In fungi and algae, dicaryotic cells as well as crenocytic and polycrenocytic structures are common.

DNA-bound protëins can be divided into two categories: **histones and non-histone proteins**. Histones are the most abundant, their quanta being equivalent to that of DNA (representing up to 108 copies per cell). They are basic protëines that neutralise the ^gative charges of DNA. There are many different types, not all of which are present in eukaryotic microbial organisms.

The non-histone protëines comprise an ipirticular class: the high-mobility protëines (HMPs). These protëines, which can recognise specific regions of DNA, are involved in regulatory processes.

The basic **structural element of** the chromosome is **the nucleosome**, which is made up of a double helix of DNA wrapped around an octamëer core of histones (nucteosomal histones H2A, H2B, H3 and H4). The core particles are separated by a zone of DNA spacer. The unit particles are organised into stacks with the possible intervention of another histone (H1) which is not present in all microorganisms. Together they form the basic fibre, whose diameter is around 10 nm.

Nucteosomes are arranged in phase with the DNA sequence of a given cell type. It appears that nucteosomes are dëplacëed at the time of transcription. The basic 10 nm fibre is wound into a hëlice to give the 30 nm fibre (6 to 7 nucteosomes per turn) **" Solenoid".** The 30 nm fibre then folds into loops (superlK'lice) to give the chromatid, whose diameter varies from 300 to 500 nm. **A chromosome** results from the evolution of a chromatid filament: the chromatid curls up and sometimes undergoes a new spiral winding (figure 6).

In the centromëric region, the nucteosomes are replaced by a protëic particle 200 nm in diameter which serves as an attachment to the microtubules: **the kinetochore.** At the end of the chromatids **is the telomeric region** {TEL). The role of the tëlomëre is to stabilise the end of the linker DNA. The extremity of the tëlomëre consists of a
sërie de courtes sëquences rëpëtitives : In yeast, this extremity represents a few hundred bases (generally 300 bp) with the tandem repetition of the CXA sequence.

The chromosome of eukaryotic organisms is characterised by a non-continuous sequence of genes. The genes or groups of genes coding for proteins are interrupted or separated by non-coding sequences **called introns**; the coding sequences are called **exons** and contain 200 to 300 base pairs.

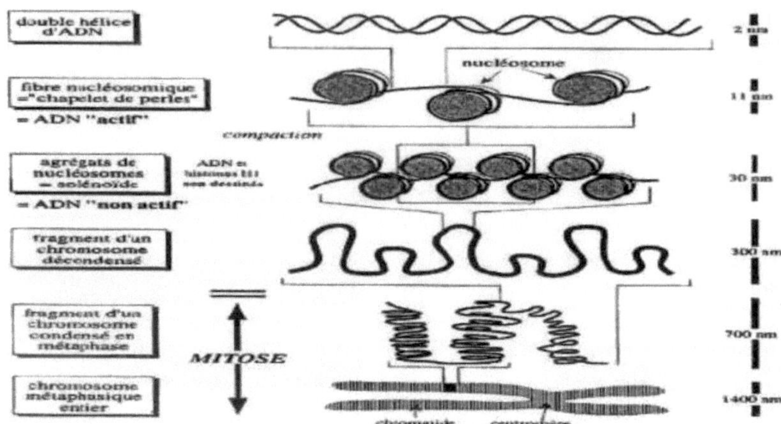

Fig 7 : Different levels of chromatin packaging

Alongside single-copy gene sequences, chromosomes contain low-frequency (103-105 copies) or high-frequency (106-107 copies) repetitive sequences. The main genes are single-copy; the low-frequency repetitive regions code for histones, ribosomal RNA and transfer RNA; they may represent up to 30% of the DNA. Highly repetitive sequences of only 5-500 bp have a structural role and are never transcribed. Some of these repetitive sequences form what are known as satellite DNAs.

There are great differences in organisation within eukaryotic microorganisms. In yeast, and particularly in *Saccharomyces cerevisiae, there* are few rëpëtiactive regions: 95% of the дёпоте are found a the simple copy state. There are no high rëpëйlr^llë regions, only medium and low rëpëtitivitë areas.These zones include the transposable ëlëments Ty, o and i), the tëlomëres, some gënes encoding protëines and those encoding RNAs (about 360 gënes of tRNA, 100 gënes of rRNA...).

When there is only one copy of each chromosome, the cell is said to be **haploid** (n chromosomes). When there are two competent sëries of chromosomes the cell is said to be **diploid** (2n chromosomes); chromosomes of the same kind are said to be **homologous.** During the life cycle of an organism, there may be a transition from haploidy to diploidy and vice versa through sexual reproduction or related mechanisms.

When the number of chromosomes is greater than two there is **polyploidy** (xn chromosomes). When the number of chromosomes is not a multiple of n there is **aneuploidy.** This can occur by the gain (hyperploidy or polysomy e.g. 2n+l) or by (hypoploidy or oligosomy: 2n-2) of one or more chromosomes.

1.4 Extrachromosomal genetic material

There is gëriel matënëtic **outside** the genophore or chromosomes. This matëriel can represent 0.1 to 1 and sometimes more than *10%* of total cellular DNA.

In eukaryotic cells, DNA is found in the mitochondria and chloroplasts. DNA, and sometimes RNA, known as "cytoplasmic" DNA, can also be found in prokaryotic or eukaryotic cells: these are more or less structured particles such as plasmids, fertility factors, "killer" factors,...

The term "**variable genetic elements**" (VGE) is used to describe all the extrachromosomal ëlë elements that can differ from one biotype to another: mainly plasmids and prophages, but also various cytoplasmic determinants.

1.4.1 Organelle DNA

Organelle DNA is double-stranded and generally circular (except for the mitochondrial DNA of certain ciliate protozoa, which is linear).

Mitochondrial DNA is wrapped around itself and bound to the inner membrane of the mitochondrion. It is around 20 pm in size in yeast (75 to 84 kb depending on the strain in *Saccharomyces cerevisiae).* There are usually several copies of this genome per organelle. There are up to 4 copies in *Saccharomyces cerevisiae*, where each cell can contain around twenty mitochondria, which on average represents 30 to 50 copies per cell, or around 15% of the total cellular DNA. The situation is very similar in moulds (*Aspergillus nidulans*).

Chloroplast DNA is a circular duplex of 100 to 200 kb: in pliotosvntlietic eukaryotic protists the number of copies of this gënome per organelle can be very ëкyë (up to 100 in *Chlamydomonas reinhardii*) whereas the number of organelles per cell is often low (sometimes just one as in *Chlamydomonas reinhardii*, whereas there are around fifteen in *Euglena gracilis*).

1.4.2- Plasmids.

Plasmids are homogenous extrachromosomal structures made up of circular double-stranded DNA of well-defined molecular size and mass (1,106 to 4,108 Da, i.e. approximately 1 to 400 kb). They are mainly found in bacteria, but also in eukaryotic microorganisms (particularly yeast).

1.4.2.1Bacterial plasmids

Bayerian plasmids replicate in a variable manner in relation to the chromosome. Plasmids can be present in multiple copies in a cell. These copies may be few ("**stringent plasmid**": 1 to 3 copies) or very numerous ("**relaxed plasmid**": up to more than 50 copies): The first case often corresponds to a dependence on the protëic replication systems of 1 host and in particular on DNA polvmërase III. In the second case, replication is independent of these systems.

A plasmid consists of **a single replicon** *(ori* sequence in *Escherichia coli*) and replication is **bidirectional** and very rapid. Plasmids are transmitted from cell to cell using two types of mechanism:Some **plasmids are conjugative,** i.e. they are transmitted during a conjugation process between two cells (contact without fusion).Other plasmids are of the **non-conjugative type**: they can only be transmitted by a transformation process.

Some plasmids can, under certain conditions, integrate into the chromosome, as some ba^riophages do: Among the conjugative plasmids of this type, the best known is fertile factor F *from Escherichia coli* which, in addition to the sexual factor, has the propertyëtë of integrating with the chromosome and inducing its transfer from one cell to another during the conjugation phase.

Plasmids carry gënes (cryptic: genes that do not give dëtectaЬle phënotypic expression) responsible for different phënotypes:

Production of toxins (hemolysin, enterotoxin, etc.), virulence factors, tumour-inducing factors in plants (e.g. *yop* genes involved in the virulence of *Yersinia enterolytica*, etc.). In *Agrobacterium, a* plant-pathogenic bacterium, the Ti and Ri plasmids carry around twenty *virulence* genes. The *Ti* part of *Agrobacteriu ..etc*

Production of bacteriocins or antibiotics: colicin (pColEl plasmid *from Escherichia coli*), pyocin (in certain Pseudomonas), vibriocins, etc.

Resistance to antibiotics or other agents (R factors or r determinants): resistance to penicillin, chloramphenicol, sulphonamides, heavy metals, toxic cations, radiation, phages, etc. Such phenomena are well known in Staphylococcus.

General metabolic properties: H2S production, citrate utilisation (in *Escherichia coli*), degradation of aromatic compounds (in Pseudomonas), plasmid pSH71 of *Streptococcus lactis*, involved in lactose catabolism, proteolytic activity...

There are incompatibilities between different plasmids (both natural and artificial): the presence of one prevents the other from becoming established (e.g. 4 classes of incompatibility in *Staphylococcus aureus).* Plasmids that are capable of transferring within a wide variety of Gram- bacterial species are called promiscuous plas**mids** (e.g. groups P, Q and W (IncP, IncQ)).

1.4.2.2Eukaryotic plasmids

Eukaryotic plasmids are few in number and are mainly found in yeast. The best known and best studied is plasmid 2g from the yeast *Saccharomyces cerevisiae:* this circular double-stranded DNA plasmid is present at a rate of 50 to 100 copies per cell (representing 3 to 5% of total cellular DNA) and is essentially localised in the nucleus: replication and transcription take place in the nucleus. It contains a single origin of replication (*ARS)* and 4 genes (or reading frames) involved in recombination and the generation of the 2 forms of the plasmid (*FLP*) and in the control of amplification and stability (*REP* 1, *REP2,* D).

Plasmids related to plasmid 2ц have ële dĕcouvered in other yeast species including *Zygosaccharomyces baillii* (pSBl, pSB2), *Zygosaccharomyces bisporus* (pSB3), *Zygosaccharomyces rouxii* (pSRl), *Kluyveromyces drosophilarum* (pKDl), *Saccharomyces italicus, Schizosaccharomyces pombe...* Although similar in size and organisation to the *Saccharomyces cerevisiae* plasmid, *the Zygosaccharomyces* and *Schizosaccharomyces* plasmids do not share any sĕquence homology with it.

1.4.3 The "killer" factor

1.4.3.1 Yeast "killer" factor.

The killer trait is the ability of certain yeasts to release into the environment a substance that is toxic to other yeasts belonging to their own species or to different species and genera. The killer characteristic exists in strains belonging to numerous species and in particular *Saccharomyces cerevisiae.*

The killer characteristic is Hë due to the presence in the cytoplasm of bicatëary RNA particles encapsulated in capsids (virus like particles: VLPs), which are intermediary between plasmids and bacteriophage viruses (they are sometimes called mycoviruses or zymophages).

There are two different types of particle living together in the cell: in *Saccharomyces cerevisiae*, the smallest particle, or M particle, and the largest particle, or L particle.

The M particle has a molecular mass close to 1,106 Da (1.9 kb) and codes for the toxic protein responsible for the killer trait: this is a dimëric (o/p) protein of 18.5 kDa which is excreted by the cell **(K+ trait). The** M particle is also responsible for killer cell immunity **(R+ trait): there are 100 to 1000 of these per cell.** The M particle is also responsible for the immunity of killer cells (**R+ characteristic**): there are between 100 and 1000 of these per cell.the largest of the particles or L particle has a molecular mass close to $2.5.10^6$ Da (4.7 kb), with slight differences depending on the killer group, and codes for the synthesis of the capsid, which is identical for both types. There are about fifteen per cell (the capsid is not essential for expression of the trait, but the large particle is essential for **multiplication** and **maintenance of** the **small** particle in the cell).

In *Kluyveromyces lactis*, the killer trait is Hë to the presence of two "linear plasmids" formedë of bicatëary DNA (50 to 100 copies per cell):**Plasmid pGKL 1**(k l) of 5.3.106 Da (8.9

kb) codes for a trimeric glycoproteic toxin (o/p/Y) and for immunity, while **plasmid pGKL2** (k2) of 8.3.106 Da (13.4 kb) is required for replication and maintenance of the system and has no capsid. -

1.4.3.2 . - Paramecies "killer" factor

There is a killer factor in paramecium: this factor controls the production of a toxic substance called **paramecin**, which is active on sensitive strains. The character depends on the presence of an infectious element k (or K). There are around 1,600 k particles per cell.

1.5 Genetic material for viruses (bacteriophages)

The genetic material of viruses (called the genophore and sometimes referred to as the chromosome) is made up of DNA or RNA and is organised in different ways. The virus genophore contains the genetic information needed to synthesise viral proteins and the enzymes required for replication.

In bacteriophages, viral genes are generally **classified** into **3 categories**:

-Early or semi-early genes are responsible for inducing the virus' reproductive cycle (destruction of host DNA, replication of viral genetic material). They are expressed very soon after infection.

-The late genes are responsible for the synthesis of capsids and other phagic structures (tail, plate, etc.) as well as lysis (synthesis of lysis proteins).

-A final group includes genes involved in the integration, excision and recombination of nucleic acids: they are only expressed in certain circumstances.

It should be noted that for some viruses, release from the infected cell does not lead to lysis: this is the case for some bacteriophages, such as M13*, in which case we talk about viruses with a parasitic cycle.

Depending on the size of their genetic make-up, viruses rely to a greater or lesser extent on the enzymes of the host cell, but in all cases this need exists, which prevents autonomous replication and means that a cell must be infected to ensure reproduction.

You can find

Linear double-stranded DNA phages :e.g. bacteriophages T1,2,3,4,5,6,7,^ (combined symmetry urophages),MU 1(u 1)..*from Escherichia coli...*

Circular double-stranded DNA phages :exp -The bacteriophage PM2 from Pseudomonas

Circular single-stranded DNA phages :exp -The bacteriophages ϕ X174, M13,f l, fd...*from Escherichia coli.*

Single-stranded RNA phages (positive RNA): e.g. *E coli* bacteriophages MS2, f2, M12, fr... (there are two cases: RNA is said to be positive if it can be directly translated: it has messenger value and is directly infectious. On the other hand, RNA is said to be negative if its complement is the messenger.

Double-stranded RNA phages: exp-bacteriophage ϕ 6, which also has a lipid envelope.

2-DNA replication

Replication (or duplication) is the mechanism that enables the reproduction of DNA molecules, which, except in special cases (single-stranded DNA viruses), are organised in a double helix. This structure is opened up, forming a "replication fork" and polymerisation of new chains of nucteotides that complement the existing chains. Replication is **semi-conservative** (Experience by Messelson and Stahl, 1957) and polymerisation, which is carried out by enzymes called **DNA polymerases**, cannot take place from a simple mixture of nucleotides.

There are several types of enzyme. Some are involved only in repair mechanisms: these are slow synthesis rate polymerases; the others, which are involved in replication itself, are fast synthesis rate polymerases.

2.1 Replication in prokaryotes

2.1.1 DNA polymerase

There are **3 DNA** polymerases in *Escherichia coli*:

DNA polymerase I: is the most abundant (95%): it is mainly involved in DNA repair but also plays a small role in replication. It polymerises **in the 5'-> 3' direction** at low speed (670 nucleotides/min).It has 3'-> 5' (the largest fragment; Klenow fragement) and 5 -> 3' (the smallest fragment) exonuclease activities and phosphorolysis and phosphorus exchange activities: one enzyme molecule is capable of polymerising up to 200 nucleotides (high processivity); its molecular mass is 109 kDa and it is bound to a Zn+ ion.

LADN polymerase II: has a controversial role: it has polymerase and exonuclease activities (3' -> 5'). it works at low speed (33 nucleotides/min); its molecular mass is 110 kDa.

DNA polymerase III (Enzyme creur):Is very active in the replication process^ It is a multimeric enzyme (subunits a, e,$,Y,[£] Д г,) in which two core molecules (each formed by three subunits "a(dnaE or polC gene) ,£ (dnaQ or mutD gene),0(holE gene)") can be linked to form a complex by a dimer of the tau (г) subunit. The catalytic activity of the enzyme is increased by the association of 4 beta (в) subunits.

Its molecular mass varies according to the organisation of the different sub-units (often around 250 kDa). This enzyme polymerises at **high speed** (104 nucleotides/min) in the **5'-> 3' direction** and has both 3'-> 5' and 5'-> 3' exonuclease activities. **NB:**

Although the depolymerase activity of DNA polymerases is not their main activity, this activity is important for the reliability of transcription, particularly in prokaryotic organisms. Thanks to their exonuclease activity, bacterial polymerases are capable **of self-correction (proofreading repair)**, which explains the **fidelity of** the copy: on average, only one error per 109 base pairs.

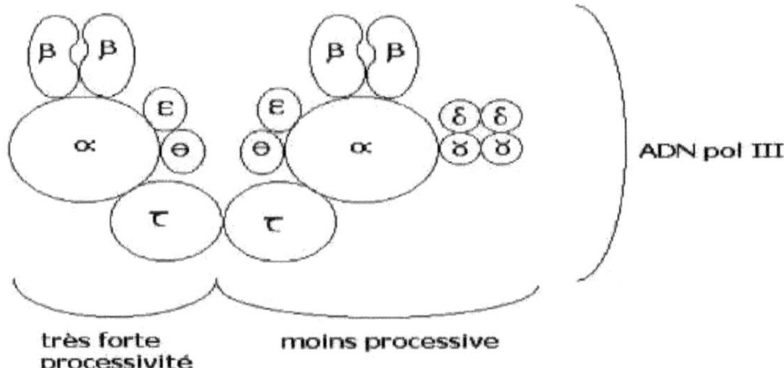

Fig8 :Simplified structure of pol III DNA

2.2.2 Mechanism of replication at *Escherichia coli*

Replication involves a large number of proteins, including a multiprotein complex known as the **replisome**.

a) Initiation: Replication begins with the opening of the double helix at a specific site (the **initiation point *ori C*).** At this point there is a sequence rich in AT nucleotides, which results in less cohesion between the chains. This opening is due to the action of a **helicase** (DNA

unwinding enzyme: DUE) which is ATP-dependent: the most important are the **rep protein** and helicase II (orIII).
This process is facilitated upstream by the binding of 4 DnaA tetramers to the DNA: this induces the formation of a bundled structure with the consequent sëparation of the strands and the accessibility of other factors. The separated strands are stabilised by proteins called "single strand binding protein" (**SSBP or** SSB=HDP: helix destabilising protein=DBP: DNA binding protein): these are also tetramers.
In addition to chain separation by helicase, replication requires unwinding of the helix structures and modification of the supercoil. This unwinding actually takes place in short bursts, thanks to the action of a **topoisomerase** (nicking closing enzyme: NCE), a subunit of gyrase, or w protein. This enzyme creates a transient break in the chain, allowing the helix to rotate (Nb; due to the 5'-> 3' specificity of polymerisation, DNA synthesis via DNA polymerase is oriented: it can only take place at one 3'OH end).
b)Elongation: The complex containing the primase and the proteins DnaB, DnaC, n, n', n" and i is called the **primosome** (there are 6 DnaC proteins for one DnaB hexamer, making a total of at least 17 protein elements per primosome).**The primase begins synthesis** by polymerising a short strand (around 30 nucleotides) of RNA whose sequence is complementary to that of the model DNA: this strand is called the initiator or primer. Once priming has taken place, DNA polymerase can then add nucleotides to **the 3' end of** the neoformed strand. **DNA polymerase III** does most of the elongation work, removing the RNA initiator strand and completing the missing part: DNA polymerase I appears to play an essential role in this phase, through both its exonuclease and polymerase activities.
To enable the two chains to grow simultaneously, synthesis is necessarily **discontinuous** on one of them, whereas it **is continuous** on the other. Replication therefore occurs **asymmetrically** on both chains.
In discontinuous growth, synthesis takes place in the permitted direction (5'-> 3') by short strands called **Okazaki fragments**: strands which are then linked by a **DNA ligase** (DNA ligase enables the double-stranded DNA to be reassembled in the case of a gap-free cut between the 5'phosphate end and the 3'OH end).
The chain synthesised continuously is called **the** leading str**and** and the other the lagging **strand**. The presence of **a loop** on the lagging strand (where polymerisation is discontinuous) allows **the replisome** to function in the same direction for both strands at the same time (fig 8).
c) Termination :
The two replication forks of the E. *coli* circular chromosome meet in a terminal region containing multiple copies of a 20-base-pair sëquence **called 'Ter'** that is diamëtrally opposite the replication initiation site.
Ter" sequences are attachment sites for a protëin called "Tus" **(terminus utilization substance).** Only one "Ter-Tus" complex functions per replication cycle. Replication is controlled by several regulatory genes: the product of the dnaA gene negatively regulates transcription of the initiator and its own synthesis during cell growth, and its level decreases, allowing replication to begin. The products of the dnaB and dnaC genes are activators of the initiation of DNA sequencing.

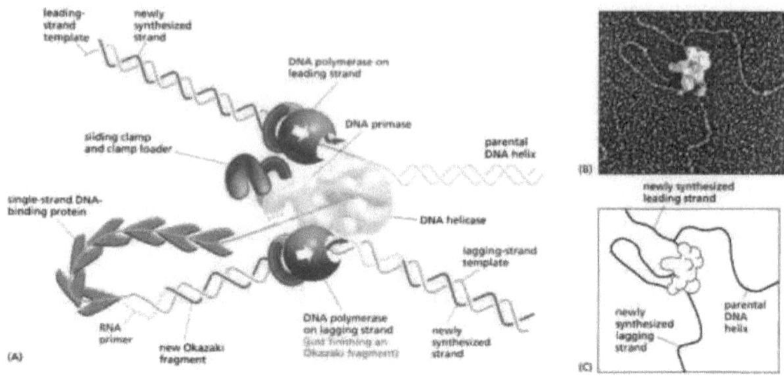

Fig 9 : Mechanism of elongation of the newly synthesised DNA chain by the DNA polvmerase III dimer in the early and late strands.

2.2 Replication in eukaryotes

2.2.1 DNA polymerases

There are also several DNA polvmërases in eukarvote microorganisms.

Table II: The main eukaryotic DNA polymerases, composition, role, activity and location

DNA polymerase	Location	Number of S/U	Prokaryotic equivalent	Activities	3'-5' exonuclease activity
Alpha (a)	core	4	DNA pol I	Initiation	-
Beta (B)	core	1		Repair and finishing	-
Gamma (Y)	mitochondria	1		Mitochondrial DNA synthesis and repair	+
Delta (6)	core	2	DNA pol III	Synthesis and finishing	+
Epsilon (E)	core	2	DNA pol II	repair	+

2.2.2 Mechanism of replication in eukaryotes

In eukaryotes, during the cell cycle, the amount of DNA doubles during the S phase, which precedes mitosis. This doubling of the genetic material, known as replication, has the same characteristics as DNA replication in prokaryotes.

a) Initiation

In yeast, the origins of replication are called replicators or **ARS** (autonomous replicative sequences), and there are around 400 replicators distributed among the 17 chromosomes of the yeast haploid genome.

The initiation of replication in all eukaryotes requires a multi-subunit protein **called ORC** (origin recognition complex) which associates with several replicator sequences.

The primer is initially created by **the primase**, which synthesises a small RNA primer of around 10 nucleotides**. The DNA polymerase** then **continues to** condense 20 desoxyribonucleotides to form a mixed strand of around 30 nucleotides, constituting the initiation point for **the activity of DNA polymerase 6**.

b) Elongation

Synthesis is taken over by **DNA pol 6**, which is **the main enzyme** for polymerising new strands complementary to the template strands. At the same time, this enzyme extends the neosynthetised strands (continuous and discontinuous) in the 5'-3' direction.

DNA pol 6 works in conjunction with **the protein PCNA** (Proliferating Cell Nuclear Antigen) which confers its attachment to the template strand for as long as possible before detaching (PCNA plays the same role as a sub ивЬё в of DNA pol III).Two other protëin complexes also function in eukaryotic cell DNA replication:**RPA**(Replication protein A) is a prot&ne that attaches to single-stranded DNA during replication and therefore has a function ëquivalent to that of the *E.coli* protëin SSB and **RFC** (Replication factor C) functions as a clamp that hooks onto PCNA and facilitates the assembly of the active replication complex.
On the other hand, on the discontinuous strand, Okazaki fragments are shorter than in prokaryotic organisms (around 135 nuckotides) because the replication forks advance more slowly. The RNA primer is progressively degraded by the combined action of RNAse H1 and nuckase Fenl. The dDNA polymërase/PCNA complex fills in the gaps and DNA ligase I binds the fragments together.
. In eukaryotes, once the DNA has been duplicated it is rëorganisedë in nuckosomes to return to its initial conformation (chromatin). DNA replication in eukaryotic organisms is accompanied by a process of histone duplication. Replication is bidirectional and occurs generally from several origins

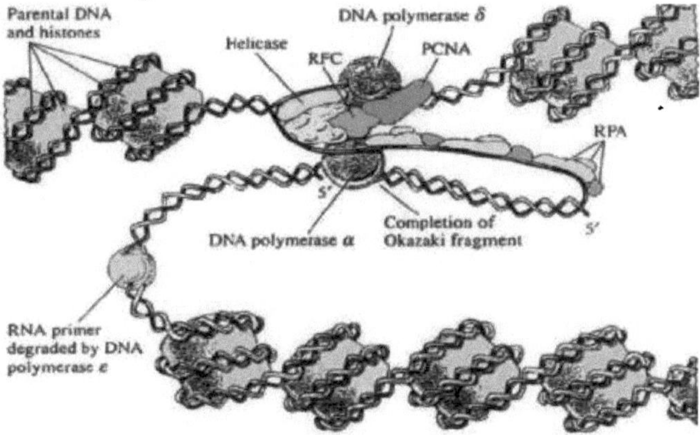

Fig 10: Mëreplication mechanism in eukaryotes

2.3 Replication of extrachromosomal DNA

Mitochondrial DNA, plasmid DNA, and in a дё^гакre way extrachromosomal DNA, are folded according to a mëcanism close to that of chromosomal DNA. These DNAs often comprise **a single replicon**.

Replication of plasmids is generally carried out according to the classic sclk'ina.

The replication of yeast plasmid 2ц, which occurs only during the S phase, leads to the formation of circular concatënaires (with form 9) which are resolved by recombination (Futcher model e)(fig 10a) whereas the replication of circular mitochondrial DNA may involve a particular structure known as **D-loop** (Fig 10.b).

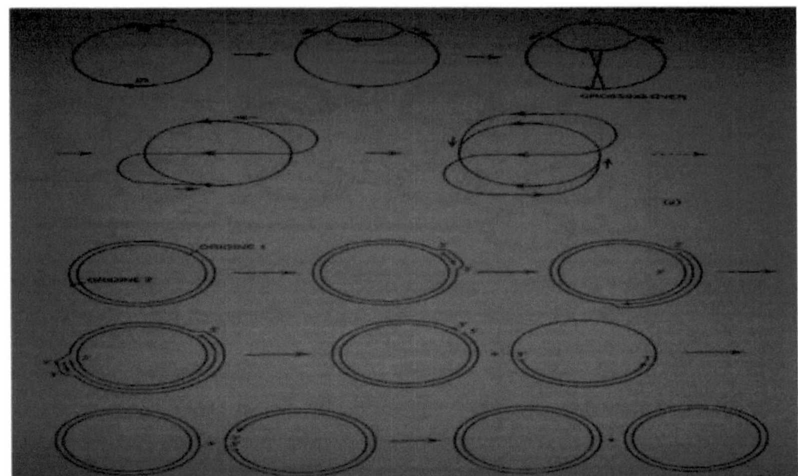

Fig 11 Cytoplasmic DNA replication

In *Saccharomyces cerevisiae*, mitochondrial DNA synthesis occurs throughout the cell cycle and is coupled to growth: this DNA appears to comprise several replication origins that do not contain an ARS sëquence.

2.4 Replication of the virus genophore

Replication of gëtic material and intracellular multiplication of viruses largely use the host's enzymatic ëequipments. The mëchanism differs according to the nature of 1 DNA or RNA.

a) Viral **DNA** is either linear or circular. The replication of circular bicatenated DNA can be similar to that of bacterial DNA with the formation **of reil or loop** replication structures (structure 9): this is the case for the early lytic replication of bacteriophage fy. After circularisation in the infected cell .

Another widespread process also concerns circularised DNA: replication by the **rolling circle** mechanism, which produces chains of genophores or concatenaries that are then split into DNA molecules before encapsidation, which takes place as the molecules are measured.

Several other replication mechanisms for linear DNA are found in the viral world:

-Replication by random strand displacement with the intervention of a protein, e.g. adenovirus

-Replication by the "hairpin" mechanism, e.g. vaccinia virus

b) Replication of both linear and circular **single-stranded DNA** requires passage through a replicative double-stranded (RF) form. In bacteriophages, single-stranded DNA is usually of the + type, i.e. it has the same polarity and sequence as the corresponding mRNA:

From the linear double-stranded replicative form by the strand displacement system: bacteriophage ф29

From the circular double-stranded replicative form by a rolling circle mechanism without concatenate formation: bacteriophage фX174

In filamentous phages of this type (M13, fd...) the size of the genophore can vary enormously because there is no capsid to limit it: it is the length of the DNA molecule that determines the size of the phage.

Replication is initiated by a primer, which in most cases is RNA or, more rarely, a protein.

The RNA primer is synthesised by the RNA polymerase (bacteriophage M 13) or by the primase directly (bacteriophage G4) or by the primase, with the host's complete mechanism and machinery, involving the primosome (bacteriophage ɸX174).

In all cases, elongation is achieved by DNA polymerase III.

c) For RNA bacteriophages:The replication of the various types of viral RNA uses different mechanisms:Double-stranded RNA viruses (e.g. reoviruses) have RNA replicases which function in a similar way to DNA viruses of the same type.

Some single-stranded RNA viruses produce a replicative form of x2 RNA either from a + strand, such as bacteriophages (ɸЯ17. QB...) or certain animal viruses (Picomavirus...), or from a - strand, such as the SV vësicular stomatitis virus. .

Lastly, certain single-stranded RNA viruses (retroviruses) transcribe RNA directly into DNA using a reverse transcriptase, and then bind a molecule of double-stranded DNA which enables transcription into RNA.

Chapter II

Expression of genetic information and its control

1-The different types of RNA

1-1-)Ribonucleic acids

The (RNA) are composedës of sëinary sequences of nucleotides with some changes compared to the тоlёсиle of DNA: ribose instead of desoxyribose and U instead of T.We distinguish :

Functional A-RNAs: **rRNA** (82%) **(ribosomal), tRNA** (16%) **(transfer), snRNAs less than 1%.**

b) Informational RNA: 2% **mRNA (messenger)**

I-2) Structural and functional properties of different types of RNA

I-2-1) Ribosomal RNAs

They are transcribed in the nucleolus and are necessary for the synthesis of protëines by binding to mRNA. They are found either attached to the endoplasmic reticulum or free in the cytoplasm.

-In procarvotes : Ribosome 70S[S/U; 50S: "2 rRNA(5S and 23S)+34 r Proteins" and S/U 30S "1 rRNA (16S) + 21 r Proteins"].

-In eucarvotes : Ribosome 80S[S/U ; 60S : " 3 rRNA (5S;5,8S et 82S) +45 r" Proteins and S/U 40S "1 rRNA(18S) + 33 r" Proteins]

I-2-2) Transfer RNA (tRNA)

They transfer and transport amino acids from the cytoplasm to the ribosome, where protëines are synthesised during translation. There is at least one tRNA for each of the 20 amino acids. Each cell contains around 100,000 copies of tRNA. The secondary structure of tRNAs is stable (a few hours to a few weeks). There are several characteristic regions common to all tRNAs.

J I .'extreme 5'phosphate in double-stranded form whereas i'extremife 3'OH is formed from 04 single-stranded nucleotides which has the sequence XCCA (X: any base)and whose adënine is the amino acid attachment site.

-*S* tRNAs contain several modified bases in particular regions: dihydrouridine (DHU), ribosylthymine (rT), pseudouridine (T), and inosine (I).

J tRNAs have **three single-stranded loops**: (1)- The **anti-codon** loop. (2)- The DHU or D loop.(3)- **The T'I'C loop** which contains the TTC (thymidine- pseudouridine-cytosine) sëquence.

I-2-3) Messenger RNA (mRNA)

These are high molecular weight RNAs whose base sequence is complementary to the DNA that gave rise to them. The lifespan of mRNAs is very short, with a low turnover rate and rapid degradation. Formed from a single nucleotide chain, a group of 3 nucleotides on the mRNA forms a code corresponding to an amino acid.

I-2-4) smal nuclear RNAs

snRNAs are specific to eukaryotic nuclei. They are heterogeneous, vary greatly in size and are involved in the maturation of pre-mRNAs into mature mRNAs.

2) Transcription

This is the process by which RNA is synthesised from a DNA template, from which it generates (mRNA), (tRNA), (rRNA) and other RNA molecules that have a structural and catalytic function. RNA polymerases are the enzymes responsible for transcription. RNA is always synthesised from DNA in the 5'-3' direction, anti-parallel to the "anti-sense" template strand and also in a complementary fashion (G/C and A/U pairings). A nucleotide is added to the free 3' OH end of the RNA strand being synthesised. Also bearing in mind that Mg^{2+} ions are essential for RNA polymerase to function.

In prokaryotes, a single enzyme catalyses the synthesis of all three classes of RNA. In eukaryotes, on the other hand, there are three different types of RNA polymerase that are structurally similar to those of bacteria: RNA polymerase I: transcription of ribosomal RNAs;RNA polymerase II: transcription of messenger RNAs and snRNAs;RNA polymerase III: transcription of transfer RNAs and 5S rRNAs.

2-1) Definition of a gene

A gene is a DNA sequence containing the information required to synthesise an RNA or a protein. It consists of: a **promoter** which serves as a binding site for RNA polymerase (and transcription factors in eukaryotes) and directs transcription initiation. Just after the promoter part is **the coding part of the gene**. In higher eukaryotes, the coding part of the gene is made up of sequences **of exons (coding sequences)** and **introns (non-coding sequences)** removed after maturation of the pre-messenger RNA. Finally, the **termination sequences** provide the signal for detachment of the RNA polymerase from the template.

NB:

Genes expressed in all cells: **house keeping genes**. Genes expressed only in a few tissues or even in a single cell type = **tissue-specific genes.**

2-2) Molecular-scale transcription in prokaryotes a)l/initiation

Transcription initiation begins at a promoter sëquence, three consensus sëquences are identified: **(1)-The Pribnow box, (-12 to -7). (2)**-The **-35 sequence**, (-35 to -30). **(3)-The UP sequence** (upstream promoter), rich in AT -40 **and -60 (**highly expressed promoters). The RNA polymerase **attaches** to the promoter **in two stages**: The first step **involves weak attachment** to the -35 region of the promoter while the DNA is still double-stranded.The second step involves a change from weak attachment to **stronger attachment** between the RNA polymerase and the DNA involving local unwinding (about 17 bases) to expose a short single-stranded DNA sequence that will form the template. This opening takes place at nucleotide +1 up to about halfway down the -17 sequence. The RNA polymerase is thus positioned at the site where polymerisation begins. The presence of **the o factor** (accessory subunit) with the core enzyme (a2ββ'ю) gives the RNA polymerase a very high affinity for the promoter. After transcription has started, the o factor is removed and it is the core enzyme that continues the elongation process.

b) Elongation

Elongation is achieved by RNA polymerase by adding **NTPs to the 3' end of** the RNA chain during synthesis. As the enzyme proceeds to open the DNA double helix, single-stranded RNA copied from the template strand is released. This enzyme **needs no primer**, uses only one strand as a template and progresses at a rate of around 30-50 nucleotides/sec. It should also be noted **that there is no exonuclease activity**, which means that transcription is less faithful than replication. The error rate in RNA synthesis is 1/104 or 1/105.

c) Termination

RNA elongation continues until the **termination signals are read**, at which point the RNA polymerase releases the template DNA chain and the neosynthetised RNA chain. **Bacteria** appear **to** use **two** distinct **strategies** to terminate transcription. The first mechanism involves the formation of **a hairpin** that destabilises and disengages the RNA polymerase from the RNA:DNA hybrid after transcription of a G-C-rich palindromic sequence followed by a U-rich region.

The second mechanism involves **the Rho protein**, an enzyme with ATPase activity that allows it to move unidirectionally along the RNA until it reaches the transcription bubble. It then dissociates the DNA-RNA hybrid by cutting the hydrogen bonds at the DNA-RNA hetero-duplex, thereby completely freeing the latter from the transcription complex.

d) Post-transcriptional modifications in prokaryotes

In prokaryotes, mRNA molecules undergo no post-transcriptional modification. In fact, most are translated during transcription. However, rRNA and tRNA are produced by cleavage and modification of native RNA. For example, in *E. coli*, three types of rRNA and one type of tRNA are produced by cleavage of primary RNA, which also contains spacers. Sometimes, tRNAs are also modified by the addition of CCA nucleotides at the 3' end to tRNAs that do not yet have this end.

NB; About polycistronic RNAs Certain prokaryotic genes are assembled structures that contain several coding parts of genes, involved in the same metabolic pathway, under the control of the same promoter, these are the operons.

2-3) Molecular-scale transcription in eukaryotes

a) Initiation

RNA polymerase II is made up of 12 subunits and cannot transcribe *in vitro* without the intervention of certain proteins called basic transcription factors (TFs). Initiation begins with the binding of the protein **TBP** (TATA-binding-protein) to **the TATA box**, then the transcription factor **TFIID** also binds to the DNA on either side of TBP. Another complex, in the form of **TFIIF** and RNA polymerase **II**, associates with it.

By interacting with **TFIIB**, **TFIIF** helps RNA polymerase II find its promoters. Finally, **TFIIH** and **TFIIE** join to form a firm complex. **TFIIH** acts as **a helicase** that unwinds DNA to initiate transcription, creating an open complex, and via its kinase activity it phosphorylates the COOHt of RNA polymerase II, causing a change in the conformation of the complex.

b) Elongation : After 60-70 nucleotides of RNA, TFIIE and TFIIH leave the complex, marking the start of the elongation phase. The elongation mechanism is essentially the same in bacteria and eukaryotes. One of the main **differences** is **the duration and length** a of the transcript to be synthesised (several hours given the presence of introns despite a polymerisation rate of 2000 nucleotides/minutes.

c) Termination: Transcription termination signals **are not yet well defined**. However, there is a consensus sequence **AAUAAA** which apparently acts as a signal for an RNA-endonuclease to activate and cut the transcript 3'.

RNA polymerase II is dephosphorylated and recycled, and both the **poly(A)**-dependent pathway and the **Senl-dependent** pathway are currently adopted.

d) Maturation of mRNA in eukaryotes: The mRNA molecule resulting from transcription undergoes modifications **at both ends**. A specific nucleotide, methyl-GTP, is added at **the 5' end**, even before transcription is complete. It forms 1) **the 5' "cap" of** the mRNA (capping). When transcription is complete, **a poly-A-polymerase** catalyses the addition of **AMP** to the

3' end, up to more than 200 residues **(2-the "poly-A tail"**).
Unlike **3'-UTR** regions, 5' **Untranslated** Regions (**5'-UTR**) are more conserved between different species.
The maturation of mRNA also involves phases **3) of intron excision** and **exon splicing**, which take place in the nucleus, with the mRNA exported to the cytoplasm consisting solely of cut exons, with the help of nuclear particles consisting of proteins and **small nuclear RNAs (snRNA) or (snRNAs) (Spliceosome).**

3) Translation of mRNA in prokaryotes and eukaryotes

Translation requires the intervention of : **Ribosome** (locus of translation), **tRNAs** (adaptor between amino acids and genetic code) and **other factors** (initiation, elongation and termination). Protein synthesis is **polarised**, with ribosomes moving **in a 5' to 3' direction** on the mRNA, and synthesising the corresponding polypeptide from the NH2-terminus to the COOH-terminus.
The mRNA sequence is decoded by a group of **three nucleotides (codon)** which correspond to a particular **amino acid**. This genetic code has certain universal characteristics in both prokaryotes and eukaryotes, including **the absence of overlap, its redundancy** or degeneracy, so that the same amino acid can be encoded by several codons, and the presence of three **stop codons** (UAG, UGA or UAA) indicating the end of a coding sequence.The mRNA(codon)/tRNA(anticodon) pairing is **complementary and anti-parallel.**
Amino acid coupling to tRNA involves the covalent attachment of a free атте acid to the adënosine terminal of its 3'-end.The атте acid is attached by its acid function (COOH) to the 3'OH or 2'OH end of tRNAs by an ester bond in the presence of l'aminoacyl-tRNA syn^tase **(transferase).** First of all, the дтте acid is **attached** to l'ATP (aa + ATP -> aa-AMP + 2 Pi), then the ami^ acid is **transferred** to the tRNA (aa - AMP + tRNA - > aa - tRNA + AMP). The energy from the ester bond will be used to attach the ami^ acids to each other (peptide bonds).
NB: The Wobble is the incomplete pairing (floating) between the codon and the anti-codon during the translation of mRNAs into protëines. This is referred to as a **"fluctuating base"**: If the first base (5') of the anticodon pear by the tRNA is C ;A ;U ;G ;I(inosine); it may pair with a third base (3') of the mRNA codon:G ;U ;A or G ;C or U ; U,C or A, respectively. This dëgënërescence also implies that certain ami^s acids have more than one tRNA (**iso-acceptors**), or that certain tRNAs can pair with more than one codon. this is a phënomëne which has 3 intërëts: saving tRNA, faster protëic synthesis, protection against certain mutations.

3-1) Stages of translation

a) Initiation

Initiation of translation in prokaryotes and eukaryotes **requires** a specific tRNA initiator: **Met-tRNA^{Met}** used to incorporate the initial mëthionine residue of all protëines.In *E coli*, a spëcific form of Met-tRNA^{Met} is required for initiation: **N-formylmethionine-tRNA^{ifmet}** (fMet-tRNA^{fMet}). Formylation occurs after the mëthionine has attacked the tRNA. It is catalysed by methionyl-tRNA formyltransferase.
Met-tRNA^{Met} and fMet-tRNA^{fMet} recognise the same initiation codon: **5'- AUG-3'**.In the polycistronic RNAs of prokaryotes, the AUG codon is adjacent to a **Shine-Delgarno** ëlëment. This ëlëment is recognised by comptementar^ of sëquences in the 16S rRNA of the small subun^.
In prokaryotes, initiation occurs **in three stages**: 1) the 30S subunit associates with two

initiation factors **IF1 and IF3**. The role of the latter is to prevent association between 30S and 50S before the mRNA associates with the 30S subunit. Bayerian ribosomes contain three sites, the A or aminoacyl site, the P or peptidyl site, and the E or exit site.

fmet-tRNA **is the only** aminoacyl tRNA that binds **to the P site,** all other tRNAs bind to the A site. The E site is where uncharged tRNAs bind. The initiation factor IF1 attaches to the A site and prevents tRNA from attaching to this site during the initiation phase. 2) The complex consisting of the 30S subunit, IF-3 and mRNA is joined by IF2+GTP and fmet-$tRNA^{fmet}$. 3). This large complex combines with the 50S subunit and simultaneously, the GTP attached to IF-2 is hydrolysed to GDP and Pi, which are released from the complex.

All the initiation factors are then released from the ribosome. The completion of these steps gives us a complex called the initiation complex.

In eukaryotes, eukaryotic mRNA forms a complex with the ribosome and a number of specific proteins. Several of these proteins are thought to bind the two ends of the message. At the 3' end, the mRNA is bound by a protein called PAB or CBP or poly A binding protein. Eukaryotic cells contain at least nine initiation factors.

b) **Elongation**

In the first stage of elongation, the appropriate aminoacyl-tRNA is bound to a complex of **EF-Tu and GTP**. The complex formed binds to the A site. The GTP is hydrolysed and an **EF-Tu-GDP** complex is released from the 70S ribosome. In the second stage of elongation, a **peptide bond,** catalysed by peptidyl-transferase, is formed between the two amino acids bound by their tRNA to the A site and the P site of the ribosome, producing a dipeptidyl-tRNA in the A site, leaving the P site with an uncharged tRNAmet.

In the third stage of elongation, the ribosome moves one codon towards the 3' end of the mRNA. The uncharged tRNA moves to the E site and then into the cytosol. The movement of the ribosome **is** called **translocation** and requires **a translocase** (EF-G) which is also a GTPase. The movement is made possible by a change in the three-dimensional conformation of the whole ribosome. The structure of **EF-G** is similar to that of the **EF-Tu/tRNA** complex, suggesting that **EF-G** can associate with the A site and move tRNA peptidyl. Elongation **in eukaryotes** is very similar to that in prokaryotes. However, eukaryotic ribosomes **do not have an E site**. Uncharged tRNAs leave the P site directly.

c) Termination

The signal for termination is one of the termination codons in mRNA: UAA, UAG and UGA. **In bacteria**, three release factors, **RF1, RF2 and RF3** (or **eRF in eukaryotes**) contribute to 1) **hydrolysis of** the terminal bond between peptidyl and tRNA, 2) **release of** the free polypeptide and the last uncharged tRNA from the P site and 3) **dissociation** of the 70S ribosome into 50S and 30S. (There are 2 termination factors in prokaryotes and only one in eukaryotes). RF1 recognises the UAG and UAA termination codons, and RF2 recognises the UGA and UAA codons.

These two termination factors bind to the termination codon and induce peptidyl transferase to transfer the polypeptide to a molecule of water instead of another amino acid. These termination factors have a structure similar to that of tRNA. The specific function of the RF-3 termination factor is not yet well established, but it is thought to release the ribosomal subunit.

4-Regulation and control of the expression of genetic information

In microorganisms, the regulation of protein synthesis is highly efficient. Gene expression can be controlled at all levels between the gene and the protein: activation or modification of the gene structure, transcription, maturation of primary mRNA, nucleus-cytoplasm transport in eukaryotic organisms, translation, stabilisation of certain types of mRNA.

Most of the control of protein synthesis takes place at the transcription initiation stage: these phenomena have been particularly well studied in prokaryotic organisms.

4.1 - Control of transcription initiation in bacteria

There are transcription control units corresponding **to operons**; an operon contains a series of one to ten structural genes, each of which codes for a peptide; these genes follow one another on the chromosome and form a transcription unit that begins with an initiation sequence and ends with a termination sequence.

The initiation sequence is preceded by **a promoter**, the RNA polymerase binding site, which controls the transcription of all the genes that depend on it. The regulation of an operon generally involves an operator gene and a regulator gene. The operator gene (or site) is adjacent to the promoter, while the regulator gene (an independent transcription unit with its own promoter) enables the synthesis of a regulator protein which acts at the level of the operator, causing the RNA polymerase to bind or pass (negative action) or, on the contrary, facilitating it (positive action).

There are several types of regulation depending on various factors:

- **Specificity**: control can be specific and limited, acting only on the genes of one operon, or on the contrary non-specific, likely to affect several operons. Non-specific systems are global control systems and generally involve a common regulator.
- **Nature of the** promoter/RNA polymerase/regulatory protein **interaction** (positive or negative action).
- **Final result**: this can take two forms. Depending on the case, there is induction in catabolism or repression in anabolism.
- **NB:** the terms **"cis" and "trans" regulation** can also be used **cis regulator** corresponds to a motif present on the DNA which exerts a regulatory effect, e.g. an Operator. A locus acts in cis if it is on the same DNA molecule as its target. Most often, it is upstream of the regulated gene.**trans regulator**: ga may be a transcription factor, for example, it is not a sequence element, but ga may be a protein which acts by binding to a target element, which may be of the operator type. It can control another locus, even from another DNA molecule.

4.2 - Specific control :

4.2.1Negative specific control

a) Induction

The most classic example is the *E coli* **lactose operon** (Fig 11). It is made up of three structural genes *lacZ, lacY* and *lacA*, a *lacO* operon (28 bp) located after the *lacP* promoter and adjacent to the *lacZ* gene and finally a *lacI* regulatory gene which is close to the operon.

The enzymes required for lactose utilization, and in particular the β- galactosidase required for hydrolysis, are only synthesized in the presence of lactose (or a structural analogue).

The regulator gene normally synthetises a protein made up of 4 subunits. This protein is inhibitory in the absence of the lactose inducer. It binds to the operators and in particular to *oi* (for which it has the greatest affinity) and prevents the action of RNA polymerase, which ultimately blocks enzyme synthesis.

In the presence of lactose, small quantities of **allolactose** (an isomer of lactose in which the galactosyl group is displaced) are formed. This product combines with the regulatory protein, rendering the protein inactive by greatly reducing its affinity for the operator: the RNA polymerase can then bind and transcription and enzyme synthesis can take place.

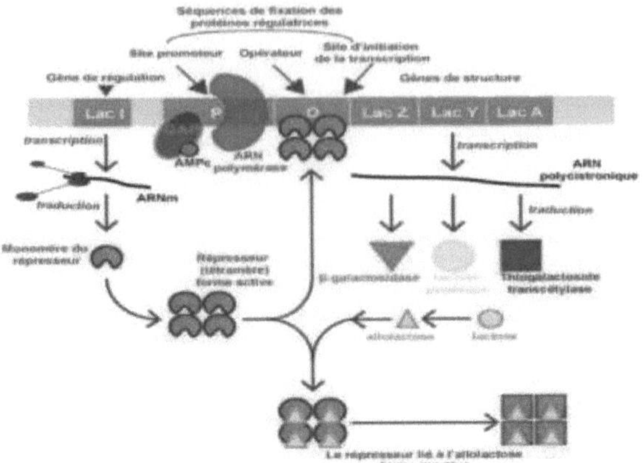

Fig 12 :The E coli lactose operon and how it works.

b) Repression

The classic example is the **threonine operon** in *E c* oli (Fig 12). **In the presence of threonine**, the synthesis of enzymes involved in the production of this amino acid, in particular threonine synthetase, is **prevented**. The regulatory gene normally synthetises a protein which is inactive in the absence of the threonine repressor.

In the presence of the repressor, a complex is formed with the protein and this complex prevents the RNA polymerase from binding. Transcription and synthesis do not take place: this is the repressed state.

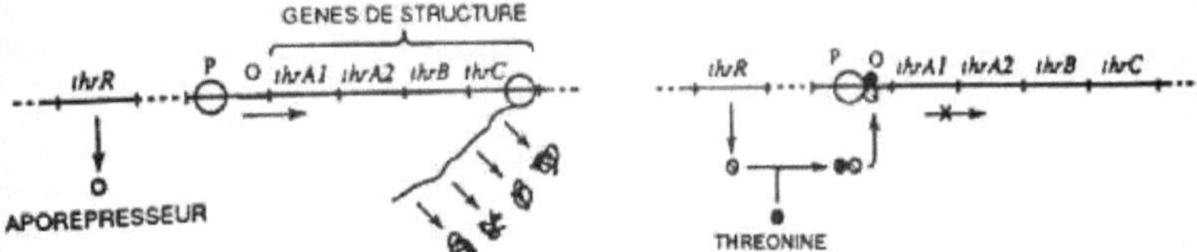

Fig 13:Function of the threonine operon

4.2 .2 Contrdle specific positive

a) Induction

An example of positive induction is provided by **the maltose operon** *of E coli* (Fig 13).the "malA" group of genes contains two structural genes which code for maltose catabolism enzymes: amylomaltase and maltodextrin phosphorylase. These enzymes are only synthesised in the presence of maltose and/or maltotriose.

Maltose acts as a coactivator for the binding of RNA polymerase in conjunction with the *malT* gene product.

In the absence of maltose, this product is no longer an activator. It should be noted that the activator product is also active on another group of closely related genes, also involved in maltose metabolism: the "*malB"* genes*: malG, F, E, K, B,* Mimplicated in transport as well as the *malS* gene.

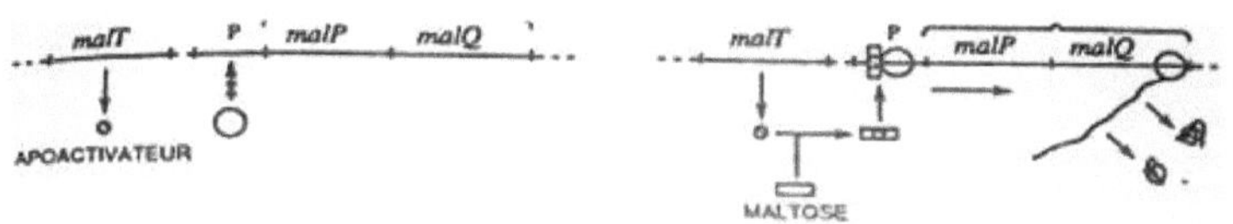

Fig 14 :Control system + maltose induction *by Escherichia coli* (malP : maltodextrin phosphorylase ; malQ : amylomaltase)

b) Repression: This occurs in the **alkaline phosphatase operon** *of E coli* (Fig 14).excess inorganic phosphate represses synthesis of alkaline phosphatase (phoA). It acts as an antiactivator of the product of the phoB (or phoR) regulatory gene, which is required for the binding of RNA polymerase. In the absence of phosphate, activation is correct and synthesis takes place.

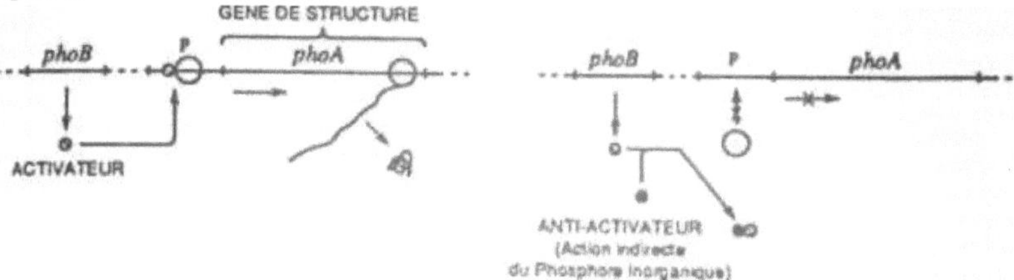

Fig 15: *Escherichia coli* phosphatase repression + control system (phoA: alkaline phosphatase)

4.3 - Contrdle non specifique

a) Catabolic suppression by glucose

The presence of a product will stop the synthesis of numerous enzymes belonging to more than one operon (Tig 15,16). This repression occurs mainly in catabolic enzyme systems which are generally inducible: hence the name catabolic repression. One of the consequences of this system is **the dialectic** phënomëne which appears in the presence of a glucose/lactose mëlange: the induction of в-galactosidase and" therefore the use of lactose cannot take place as long as glucose remains in the medium).Its mëcanism is well known in *E coli* .

In the presence of glucose, the level of cyclic AMP falls, a fall which can be attributed to several factors: 1)a drop in the ënergëtic charge of the expressCe cell 2)activation of phosphodiesterase (which degrades cAMP) and above all 3)inhibition of the function of radenyl-cyclase.**Cyclic AMP** is an essential component of this regulatory system: it acts as a **coactivator** of transcription initiation.

A protein, known as **CAP (or CRP)** and made up of two sub-units, is the main activator, but its activity requires the presence of cyclic AMP. It is therefore **a positive type of** regulatory system. 4) formation of **protein III^{glu}** which binds to the CAP molecule which then assumes the **crCAP** conformation: this complex inhibits adenyl cyclase, resulting in a drop in cAMP levels.

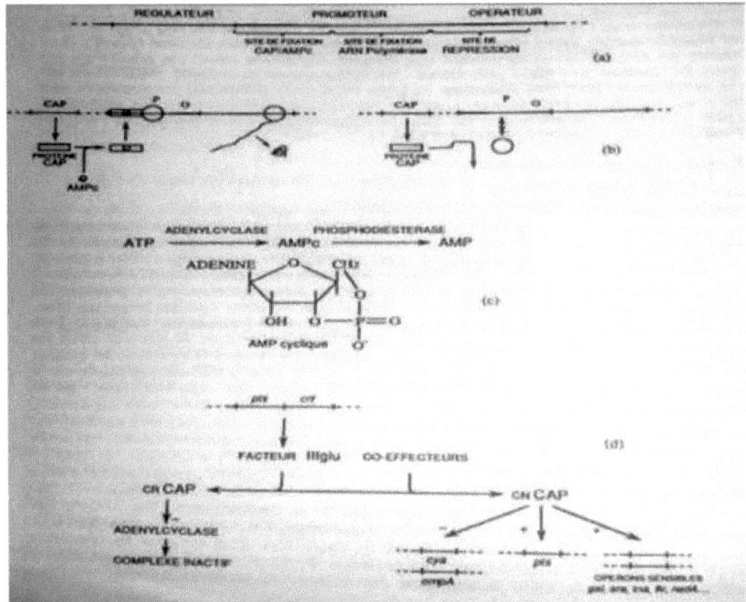

Fig 16 :_Rë catabolic **pressure** by glucose

<u>**In the absence of glucose**, there are few Glu III proteins</u> and the molecules present are phosphorylated:1) in this form they activate adenyl cyclase, resulting <u>in an increase</u> in cAMP levels. 2) In the presence of cAMP, the CAP protein assumes its **cnCAP** activating conformation. Similarly, the gluIII protein inhibits certain "non-phosphorylating" sugar transport systems such as <u>lactose permease</u>.

<u>**The Crabtree effect**</u> is <u>a particular manifestation</u> of the repressive effect of glucose. <u>Glucose inhibits the synthesis of respiratory enzymes, while at the same time inducing fermentative enzymes</u>.

This effect, first demonstrated in the cells of higher eukaryotic organisms, is present in many microorganisms. In *Escherichia coli*, glucose inhibits the synthesis of succinodeshydrogenase and activates that of glyceraldehyde-3P dehydrogenase.

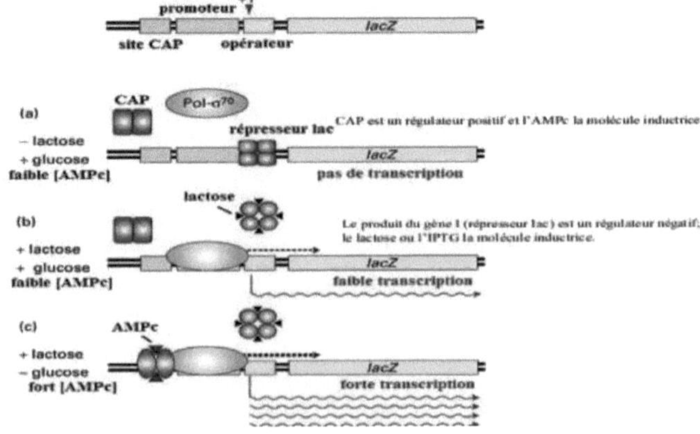

Fig 17 Recapitulation of the operation of the Lactose operon

b) Other non-specific systems

There are also cases of catabolic repression by nitrogen and more particularly **ammonium**: this repression concerns the synthesis of various nitrogen metabolism enzymes (nitrate reductase, nitrogenase, acetamidase, urease, xanthine dehydrogenase, asparaginase...).in aero-anaerobic bacteria such as *Escherichia coli* or *Salmonella typhimurium*.

There is a hiërarchisë contrdle of respiration and fermentation. Aërobic conditions induce the synthesis of at least 19 protëines (pyruvate dëshydrogënase, Krebs cycle enzymes, etc.), while anaërobic conditions induce the synthesis of at least 18 fermentative enzymes (glycolysis enzymes, pyruvate formate lyase, etc.). In *Escherichia coli*, the ***fnr* gene** positively controls the "anaërobic stimulus", while *in Salmonella typhimurium*, there are at least 2 regulatory genes.

Other more or less specific regulations are linked to this system and involve both repression and initiation of enzymatic activity: in *Escherichia coli*, **oxygen** represses nitrate reductase synthesis, and there is no ethanol dehydrogenase activity in the presence of fumarate, and so on.**Light** is another example of a factor that indirectly causes non-specific contrdle: the biogenesis of chromatophores in non-sulphurous purple bacteria is stimulated by light.

4.4 Multiple or coordinated controls

a) Inducible catabolic operators sensitive to catabolic repression by glucose

The control of **the *gal* operon** *in E coli* is complex. The enzymes involved in galactose metabolism, which are inducible enzymes (negative regulation), have their synthesis repressed by glucose, but one of the enzymes, **galactose epimerase**, has an unrepresentable 'constitutive' level of synthesis. **In the absence of galactose,**.

This enzyme is necessary for the cell to convert UDPglucose -> UDPgalactose, which is fundamental for parietal synthesis. The operon contains 3 structural genes *galE* (epimerase), *galT* (transferase) and *galK* (kinase) which are transcribed from 2 partially overlapping promoters PG1 and PG2: one (PG2) does not require the presence of the cAMP/CAP complex, whereas the other (PG1) does. **In the absence of glucose**, the CAP-AMPc complex binds with the cat site (located at -35) and PGi functions, allowing l-type transcription, while PG2 is blocked.

In the presence of glucose, only PG2 is active, allowing type 2 transcription and hence epimerase synthesis. The two promoters are negatively contrasted by the product of the *galR* regulatory gene via the OE and OI operators: this protein forms a tetramer that associates with the two operators, forming a loop structure.

b) Specific mixed regulation

Regulation of this type exists **in the arabinose operon** *of E coli* (Fig 17), which controls 3 enzymes: arabinose isomerase, ribulose kinase and ribulose phosphate isomerase.

The *ara* operon consists of the structural genes *araA, B* and *D* and a regulatory gene *araC*.The *araC* product can take two conformations:

The system is sensitive to catabolic repression: the CAP-AMPc complex activates both promoters. In its absence, and under induction conditions, there is only weak expression from PABD and no expression from Pc.

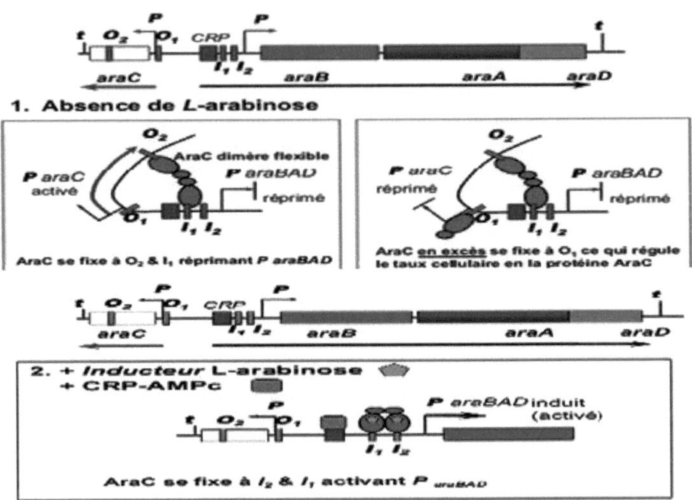

Fig 18 : Specific mixed operon arabinose regulation system of *Escherichia coli*

c) Other systems with multiple operators and promoters

Another example involving autogenous regulation can be cited. In *Escherichia coli*, the repressor **of the tryptophan operon** encoded by the trpR regulatory gene acts on 3 loci: the *trpEDCBA* operator controlling the tryptophan synthesis genes, the *aroH* operator controlling the genes of 3 enzymes catalysing the initial reaction in the synthesis of all the aromatic amino acids, and finally its own trpR operator.

The trpR gene is rëpптë by its own product. It should be noted that the sëquence of the trpE, D, C, B and A structural genes has two promoters: a promoter located^ at the dëbut and containing the opërator and an internal secondary promoter (Fig 18).

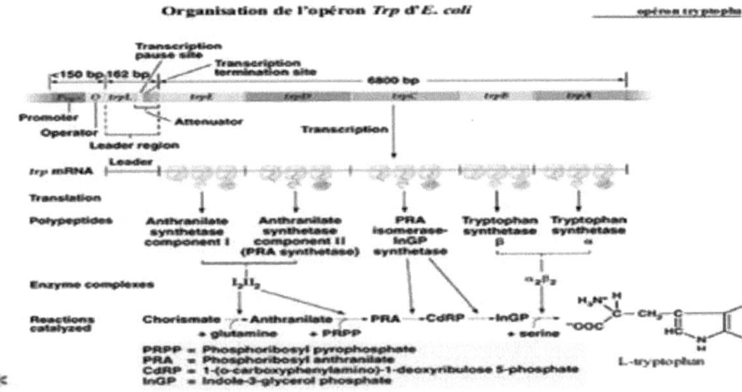

Fig 19: Tryptophan operon regulation system

NB:

The position of the ribosome plays an important role in the phënomëne of attënuation (Fig 19).

Situation I: trvptophan is absent (or in insufficient quantities)

a) the trp-tRNAs are unavailable, the ribosome stops at the trp codons, which covers region 1.

b) region 1 cannot be paired with region 2. instead, region 2 is paired with region 2. region 3.

c) region 3 cannot therefore be matched with region 4.

d) 1 RNA polymërase continues to transcribe the entire coding sëquence oc which allows complete mRNA synthesis.

Situation 2: Trvptophanc is abundant

a) the ribosome does not stop at the trp codons; it continues to translate the leader sequence, stopping at region 2

c) region 2 cannot be matched with region 3; region 3 is then matched with region 2. region 4.

d) this 3-4 pairing constitutes the "attenuator" sequence and acts as a termination signal

e) transcription is completed before the RNA polymërase reaches the gënes for tryptophan synthesis

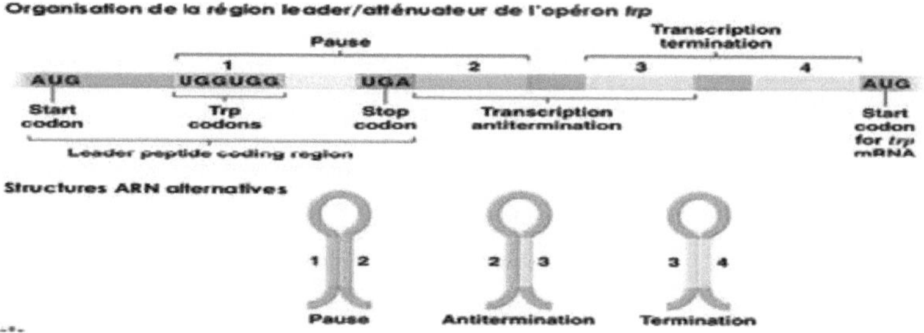

Fig 20 :Possible formation of pairing loops within the tryptophan operon

4.5Other tvpes of contrdle

a)Regulation bv factors c :

In ***Bacillus subtilis***, several factors ***a*** may be involved in the binding of RNA polymerase. Binding is heterogeneous and the core enzyme can combine with different factors.

This leads to major changes in gene expression at the time of sporulation or infection by a lytic bacteriophage (**SPOl phage**): there is a cascade action of c factors. The normal factor is **c43**: this factor is involved in the RNA polymerase that is active in vegetative cells. During sporulation, **c43** is replaced at the start of the process by the **c37** factor (and for certain genes by c32) and then by the **c29** factor. During lytic infection by bacteriophage, the c43 factor is replaced by the **gp28** factor, then by the **gp33/gp34** pair.

In this case, the c43 factor enables transcription of the early viral genes, one of which codes for the new factor that takes over transcription of the intermediate genes and so on.

In ***Escherichia coli***, the synthesis of heat shock proteins (17 proteins) which appear in response to heat treatment are under the control of c factors (**c 70** is replaced by c **32**).

In ***Klebsiella pneumoniae***, *a* bacterium that can fix atmospheric nitrogen, a similar mechanism initiates the transcription of the genes ***(nif)*** responsible for synthesising the enzymes involved: factor **o60** is involved, consecutively to a nitrogen deficiency.

A similar situation exists in *Escherichia coli* for the global regulation of nitrogen metabolism:

in the absence of ammonium, the synthesis of enzymes involved in the degradation of alternative sources is activated by the intervention of factor σ60.

b) Regulation by membrane interaction

There are regulatory systems involving the cell membrane: this is the case for the ***put*** system in ***S typhimurium***, which is involved in the degradation **of proline**. This system comprises the ***putP*** gene, which codes for proline permease, and ***theputA*** gene, which codes for proline oxidase, a membrane-bound enzyme.

The two genes are adjacent and are transcribed in opposite ways from a common promoter. The *PutA* (proline oxidase) protein is associated with the membrane as soon as it is synthesised: when all the sites are used, the surplus goes into the cytoplasm to inhibit transcription of the two genes.

C.Other transcription and translation controls in prokaryotic organisms

C.1- Transcription/translation interactions

C.1.1)Regulation by an antisense RNA: The synthesis of the **OmpF and OmpC** proteins (porins) involved in passive diffusion in *Escherichia coli* depends on the osmotic pressure of the medium: the level of OmpF decreases when the osmotic pressure increases, whereas, in the opposite case, there is an increase in this level and a parallel decrease in that of OmpC.

The *envZ* gene codes for an osmosensitive membrane protein. When the osmotic pressure of the medium increases, the product of this gene activates the **OmpR protein** (*ompR* gene). This protein positively regulates the transcription of the independent genes ompF and ompC. The **ompC** locus is in fact made up of two contiguous genes transcribed in opposite ways: ompC and **micF**. The increase in osmotic pressure therefore induces transcription of **the micF RNA**, which is known as '**antisense**' and is able to hybridize with the ompF RNA, preventing its translation.

C.2.2 Translation regulation

i) Regulation of translation initiation by the secondary structure :

Various initiation sites can be controlled by the secondary structure of the mRNA, which blocks or deëbloque them depending on the conformation. The formation of loops on the mRNA prevents translation.

ii) Regulation of translation initiation by a protein (autogenous regulation of ribosomal proteins)

This regulation is frequent for opërons encoding ribosomal protëines (opërons *Lll, S10, a, etc* in *E coli*). One of the proteins encoded by the operon (respectively proteins Ll, L4, S4, L10 etc) serves as a translation repressor at the level of the corresponding mRNA .

iii) Regulation of translation termination

This phenomenon exists in the case of autoregulation of the translation of the RF2 factor involved in termination. There is a shift of one nucleotide between the reading frame of the first 25 amino acids and that of the following ones, with the UGA codon at the end of the first frame. In the presence of an excess of RF2, this stop codon is recognised and translation is interrupted. If there is a deficit in RF2, translation continues by "skipping" the stop codon: the UGA triplet is read with the next C nucleotide in the form of a quartet which is interpreted as aspartate and the RF2 molecule is synthesised.

iv) "Stringent response

In the presence of poor conditions due to a lack of amino acids, the bacterial metabolism reacts by taking a lower level: this response is coupled with a drop in the level of stable RNA. This may be due to the absence of a particular amino acid or to a mutation in a tRNA. Normal ribosome function is blocked. Guanosine tetraphosphate ppGpp and pentaphosphate pppGpp accumulate (idling reaction).

The induction of this production is due to the binding of uncharged tRNA to the ribosome and the action **of the "stringent" factor**. This factor is a protein with pyrophosphate transferase properties, encoded by the *relA* gene: it associates with certain ribosomes (1 in 200), catalysing the synthesis of guanosine polyphosphates from GDP or GDP and ATP. ppGpp blocks GTP-dependent reactions in translation by coupling with initiation or elongation factors, and also blocks the action of RNA polymerase on stable RNA genes. ppGpp is an alarmone.

4.7 Control mechanisms in eukaryotic organisms

The basic mechanisms of regulation are similar to those found in prokaryotic organisms, but in general the systems are more complex. It should be noted that gene expression in eukaryotic organisms can be controlled by DNA methylation reactions (at the cytosine level). As in prokaryotic organisms, regulation occurs mainly at transcription initiation. **UAS or URS** sequences are involved and are recognised by regulatory proteins.

4.7.1- Transcription initiation control

Regulatory systems have been studied mainly in yeast and some moulds. Control of transcription initiation is specific to many yeast genes.

The region located upstream (5') of these genes has components that are involved in this process: a promoter site and an operator or regulator site. The control site is often an activator site: there are few examples of strictly negative regulation in yeast. The **arginine** system (**or 'operon'**) of *Saccharomyces cerevisiae* is a system integrating induction and repression mechanisms.

There is a coupling between the functioning of two genes: the biosynthesis gene and the degradation gene. The product of the regulator gene (r) **represses** ornithine transcarboxylase synthesis (anabolism) in the presence of arginine (repression).

The arginine-r complex **inactivates** the r' product of the second regulatory gene that controls ornithine transaminase synthesis (catabolism), producing an induction of this synthesis. In the absence of arginine, this synthesis is inhibited by r'.

The case of **nitrate reductase** in the mould ***Aspergillus nidulans*** *is* similar to that of arginine in yeast. In the presence of nitrate, nitrate reductase prevents the transformation of the product derived from the regulatory gene: this transformation normally causes it to change from an activating form to an inhibitory form, resulting in activation of nitrate reductase synthesis.

In the presence of ammonium, nitrate reductase allows reverse transformation and therefore inhibits its own synthesis. This is a self-regulating system.

In addition, in yeast there are genes that are controlled in a coordinated manner in the same way as bacterial operons but, more often than not, the genes concerned are not transcribed together because they are located at different sites, and even on different chromosomes. For example, 3 histidine biosynthesis genes that are regulated in a coordinated fashion are located on 3 different chromosomes.

These coordinated systems are controlled by proteins or RNA synthesised from a regulatory gene (known as an integrator) of a type similar to the bacterial regulatory gene.

The integrator gene is itself controlled by an adjacent site, the 'sensor'. This is the case for the yeast **galactose system**, which consists of a battery of genes scattered over several chromosomes, and contains perhaps one of the rare eukaryotic operons: the group of *GALI, GAL7 and GAL10* genes located on chromosome II.

Regulation of transcription of all the structural genes (at least 5) is due to the product of the *GAL4* gene. The *GAL1* and *GAL10/GAL7* genes are transcribed in opposite directions from a central region with distinct promoters.
Therefore, the two *GAL10/GAL7* genes alone can be considered to form a true operon because they form a single transcription unit. A **UAS sequence** located between the two promoters contains 4 binding sites for the *Gal4* product. **In the absence of galactose**, there is "basal" transcription of GAL1: the Gal4 product binds to site IV of the UAS sequence, but cannot activate transcription because of the simultaneous presence of the Gal 80 repressor; **in the presence of galactose**, this product dissociates and the *Gal4* product can activate transcription. **In the presence of glucose**, the *Gal4* product cannot bind and activate transcription.
The aromatic amino acid biosynthesis genes (aromatic system: arol, 2, 4, 5, 9) in ***Neurospora crassa*** constitute a pseudo-operon: in fact they are a fused gene producing a single multifunctional polypeptide.
In the same species**, the quinic acid degradation system** (*qa* genes) could be another example of a eukaryotic operon, but all its characteristics are not yet well known. **The** yeast **phosphatase system** is an example of complex regulation involving numerous factors.
Other specific regulatory mechanisms important for industrial applications need to be identified in yeast.
In ***Saccharomyces cerevisiae***, oxygen induces the synthesis of cytochrome C oxidase and peroxidase, lacticocytochrome C reductase, glycerophosphate cytochrome C reductase and other respiratory enzymes, and **stimulates mitochondrial biogenesis**.
Catabolic repression phenomena are also well known in eukaryotic organisms. **The Crabtree effect** has been studied extensively in yeast: this is a particular glucose effect that different species possess to varying degrees and which is important at industrial level. Glucose inhibits the synthesis of cytochromes and induces the synthesis of fermentation enzymes. The mechanism of this Crabtree effect appears to be **different** from that of the glucose effect on **bacterial** catabolic processes: the role of cyclic AMP is controversial. Glucose causes the appearance of secondary signals which are active on several 'sensor' type receptors, some of which are positively controlled, others negatively controlled.
In yeast, the production of heat shock proteins (which contain highly conserved amino acid sequences) is induced by an increase in temperature, as well as by other unfavourable environmental conditions: this production is controlled at transcriptional level by a positive regulatory system.

4.7.2- Other regulation mechanisms: translation control.

This control is achieved by means of nutritional or energy limitations, as well as by the correlation between the nature of the codons and the abundance of different tRNAs.
In yeast, the efficiency of ribosomes is modified as a function of growth rate and cell cycle; in the event of heat shock, there is selective translation of the mRNAs involved, due to a greater affinity for ribosomes and protein factors. In yeast, there are stringent response-type phenomena, but the precise mechanism remains obscure; similarly, there is an autoregulatory system for ribosomal protein production.
Of particular note in eukaryotic microorganisms is the role played by transposons in the control of transcription. In yeast, the most classic example is the control of sexual sign or "mating type", but there are many other cases where transposable elements are involved in modifying the expression of a gene (see chapter 4 transposition). For example, the ëlëments

Ty or 5 can be inserted at a promoter preventing transcription (for example, resulting in the His phenotype).

Chapter III

Mutations and repair systems

1-Natural or spontaneous mutations

1.1 Definition of mutation

A mutation is a hërëditary modification of the gënëtique material of an individual, in the absence of confrontation with a ëtranger gënëtique material.This modification may or may not be reflected at the phënotypic level(visiьles).

1.2 . General information on natural mutations

General information on natural mutations

-a-Spontaneity:(Lederberg experiment;velvet replica) :indëpendence on the environment: in principle a mutation occurs at random.

b-Discontinuous (abrupt): (in a single ëtape: all-or-nothing law); however, extreme behaviour, for example high-level resistance to quinolones in *E.coli*, appears following successive mutations in several genes (pleiotropic mutations).

c-Rarete :Its frequency of appearance varies between 10^{-4} and 10^{-12} but more gënërally, between 10^{-6} and 10^{-9} per gënëration

The mutation rate is the probability of a mutant appearing in a population during a given time interval. This time interval is generally the generation time for unicellular organisms.

It depends on the nature of the mutation (species, gene affected), the mutagen and its dosage (in the case of induced mutations) and possibly the surrounding physico-chemical conditions.

The mutant frequency depends on the mutation rate, the growth parameters of the wild and mutant strains and time. Its value is deduced from the slope of the curve frequence=f(n ;nbre of generations)(straight line)

d-Stability :

The trait acquired by the mutation is passed on to the offspring and is maintained in the subcultures. However, stability does not rule out the reversibility of the mutation.

Stability, however, does not rule out the reversibility of the mutation; reversibility of the mutation is possible by the intervention of a new mutation.

E-Independence and specificity: The mutation usually affects only one characteristic while respecting the others (e.g.: M.*tuberculosis* sensitive to all antibiotics ^ *M.tuberculosis* resistant to streptomycin and sensitive to all other antibiotics).

The mutation of a given trait does not alter the probability of mutation of another trait. Mutations are independent. It follows that the probability of simultaneous mutation with respect to two characteristics is ëдal e to the product of the individual probabilities.

NB:

A/Phenotypic variations: unstable, not hereditarily fixed, undergo change

1.3 Types of mutants

A-Morphological mutants:Morphological characteristics: flagella, cell dimensions, colony shape.

B-Nutritional mutants: Appearance or disappearance of a nutritional need.

C-Mutants resistant to antimicrobial agents, antibiotics and heavy metals.

D-Conditional letal mutants: Mutation in a gene whose product is essential to the cell.

E-Pleiotropic mutations: This is a mutation which affects a single gene but gives rise to several mutant phenotypes (the effect of the mutation affects different functions in the cell).

1.4 Different types of mutations

1.4.1 Classification by nature and location

Mutations can be classified **into nuclear** (or chromosomal) **mutations** and **cytoplasmic mutations** depending on whether they affect chromosomal genetic material (chromosomes or genophores) or extrachromosomal material (plasmids, mitochondria, etc.).

1.4.2 Depending on the size of the genetic material involved, a distinction is made between :

a)-Genomic mutations. These affect the number of chromosomes in eukaryotic organisms and therefore ploidy.

b)-Chromosomal mutations. They are inter- (eukaryotic organisms) or intrachromosomal (prokaryotic and eukaryotic organisms): interchromosomal mutations consist of changes between two chromosomes and intrachromosomal mutations affect the structure of a chromosome (intergenic mutations are generally referred to as chromosomal mutations).

c)-Gene mutations. These are found in both prokaryotic and eukaryotic organisms. They are mutations affecting a single gene (**intragenic mutation**): they can be more or less extensive and a distinction is made between point mutations and non-point mutations.

1.4.3- Classification according to the consequences on duplication and transcription

There are two main types of alteration:

-Mutagenic alterations resulting from a limited modification of the DNA that does not disrupt its function (generally produced by a point mutation),

-and **inactivating alterations,** caused by a mechanism that disrupts the very functioning of the genetic material (these generally have a latent effect, but a single mutation can have the same effect if it affects a vital gene).

Depending on the case, the mutation may or may not be linked to cell multiplication, and other types of alteration may be distinguished:

Alterations that do not affect duplication or transcription:-These have no consequences and are in principle non-hereditary. They involve, for example, chemical or enzymatic modification of the DNA at sites not involved in pairing (5 methylation of pyrimidine bases).

-Hereditary alterations which modify transcription but not duplication. -These are due to point mutations. Examples include alteration of the specificity of complementary base bonds by nitrous acid or mismatches caused by bromouracil.

-Transcription modification can have hidden effects or effects on the phenotype (mutation) or even lead to death.

Alterations that modify DNA duplication or cause DNA synthesis: Examples include the formation of thymine dimers by UV radiation, base alterations or loss and, in general, cases of DNA chain breaks.

Alterations modifying transcription and duplication - these are structural changes leading to major distortions in DNA chains.

1.4.4 One-off transfers

They involve a single base pair. The following cases can be distinguished **a-Substitution**

One pair of bases is replaced by another. If a purine base is replaced by another purine base or a pyrimidine base by another pyrimidine base, there is a **transition**, otherwise there is a **transversion**. In addition to the classical bases, there are natural analogues (5-methylcytosine, 5-hydroxymethyluracil, uracil) which can be incorporated. This type of mutation generally occurs during replication.

b- Deletion:-A base pair is lost. The result is a change in reading which, because of the shift produced, results in different codons: this type of mutation is known as a **frame-shift**

mutation.

c.Insertion-A base pair is added. Insertion has the same consequences as deletion on the reading of the genetic information.

d. MODIFICATION-A base pair is modified. The main modifications are the formation of tautomeric forms, cytosine deamination, depurinations, dimer formation, etc. These mutations can affect the genetic material at rest.

1.4.5 Shuffle mutations:-These involve more or less extended chromosome (or chromatid) elements ranging from a few base pairs (micro-shuffles) to polygenic fragments.-Shuffles are generally caused by breaks in DNA chains. -These breaks may occur during replication or outside it: they are most frequently observed at the time of crossovers.

Depending on whether the cuts occur before or after replication, chromosomal rearrangements can be classified into two categories: **those resulting from chromatid aberrations** and **those resulting from chromosomal alterations**. The following general cases can be distinguished.

a. Deletion:- This is the loss of a segment; when the segment is short, it is called a **microdeletion.** -Intragenic microdeletions should not be confused with point deletions as they are more extensive and therefore more difficult to reverse. -The deletion may be internal (endochromosomal) or external (terminal, exochromosomal).

b.Reversal - This is the reversal of a segment. This reversal can be consecutive to the formation of a loop. -Inversion can render a gene unfit for transcription, as it cannot be read in the correct direction.

c.Translocation-Translocation is the movement of a segment from one position to another on the same or another chromosome. If there is an exchange of segments between two non-homologous chromosomes, the translocation is said to be **reciprocal**:- it is homozygous if it affects both chromosomes of a pair at the same time and heterozygous if it affects only one of the chromosomes of a pair and one of another pair.

NB: A reciprocal translocation between two homologous sites on two homologous chromosomes is a **recombination**; on the other hand, a reciprocal translocation between non-homologous sites on any chromosome is a **mutation.**

E. Duplication

-A sequence is repeated. The result is a change in the location of the duplicated genes: this change in position can modify the expression of the gene.

1.1.6- Ploidy changes

Polyploi'die occurs following abnormal mitosis **(endomitosis)** and is then due to a non-functioning mitotic spindle. It can also occur following abnormal meiosis.

Polyploidy is hereditary. Quite common in industrial strains, it often leads to an increase in cell size and, above all, frequent failure of sexual reproduction, which hinders attempts to improve the genetics of these strains.

Aneuploidy is caused by non-disjunction of homologous chromosomes at meiosis or by abnormalities in mitosis.

1.5 - Different types of mutagenic agents (Use in mutagenesis)

There are many agents with mutagenic potential: most can be used to obtain mutations artificially, known as **induced mutations.**

1.5.1- Physical agents

a) Ultraviolet radiation

The most effective UV radiations are those with wavelengths between 200 and 300 nm, as this corresponds to the maximum absorption of nucleic acids. Low-pressure mercury

germicidal lamps are effective (^= 245 nm).
UV radiation can act directly through photochemical reactions. - The action is essentially on pyrimidine bases, resulting in **dimerisation** (thymine dimers: 2T).
The result is a deformation of the chain, the effect of which is generally lethal. This type of alteration occurs in both bacteria and eukaryotic microorganisms (yeasts, moulds).
UV rays also cause GC -> AT **transitions**, shift mutations and deletions - UV mutagenesis is a convenient method, easy to perform with simple equipment, not very dangerous for the manipulator and not very expensive. -UV mutagenesis is easily reversible due to the presence of photoreactivation phenomena alongside conventional excision repair systems.
In дёпёга1, UV radiation results in an exponential survival curve and there is no linear relationship between mutant frequency and dose.
b)Ionising radiation:-This may be **X-rays or** neutrons, dklectrons, etc. This radiation induces the formation of free radicals. -They cause chain breaks with either recombinations or point mutations occurring when the break is repaired.
In moulds, X-rays cause single and double breaks in phosphodiester bridges, as well_as dësaminations or dëshydroxylations of bases.
As the chromosomal damage caused is often very serious, these agents are only used as a last resort (in the case of organisms that are 'impermëaЬIes' to UV and not receptive to chemical mutagënes, such as certain fungal spores). - Contrary to what happens with UV rays, there is frequently a link between mutant frequency and dose.

1.5.2- Chemical agents

There are very many of them, which act by various mëcanisms. There is no дё^гак relationship between mutant frequency and dose: depending on the case, this relationship is linked or not.

A) Basic analogues

They are only effective on multiplying cells. It is often necessary to use them on an auxotrophic mutant for the base concerned.
a.1)5-Halogen uracils (5-bromouracil, 5-chlorouracil, 5-iodouracil):-These thymine analogues associate with adënine and more rarely with guanine-their action ultimately results in **an** AT -> GC **transition** (replication error) or a CG -> AT **transition** (incorporation error).
a.2)-Aminopurine-This adënine analogue pairs with pyrimidine bases inducing GC -> AT or AT > GC **transitions.**

b)- Base transformation agents

b.1)Nitrous acid (HN02):-This unstable compound mainly produces point mutations, in particular hydroxylations and oxidative dësaminations:
-Guanine -> Xanthine: this transformation is ketal
-Adënine -> Hypoxanthine: this induces an AT -> GC transition
-Cytosine -> uracil: this induces a GC -> AT transition
- This agent is practically harmless and the operation can be easily controlled: it is often used for yeasts and moulds. -Survival follows an exponential law and the mutation rate varies linearly with the dose.
b.2) Hydroxylamine and derivatives -These reducing compounds are capable of producing free radicals in the presence of oxygen and traces of metal. The products formed can inactivate DNA and cause chromosomal aberrations **-Hydroxylamine and N-methylhydroxylamine** cause point mutations, in particular cytosine -> 4 (or 4,6) Hydroxylaminecytosine transfomation, which induces the GC->AT transition.

b.3)Other producers of free radicals - There are many compounds of this type. The production of free radicals frequently depends on the presence of metals and is facilitated by radiation.
-Hydrazines and hydrazides act in the same way as hydroxylamine, their action being better at high pH levels.
-Hydrogen peroxide and organic peroxides (such as disuccinyl peroxide) act on DNA via radicals, alternating bases and causing cuts (alteration of the desoxyribose). adenine -> 7(or 8)>-N-hydroxyadinine transformation can also occur. Peroxides mainly cause inactivating alterations and produce few point mutations; organisms with a catalase are protected. Aldehydes and phenols are also capable of causing mutations.
b.4)Alkylating agents:-These are reagents capable of causing methylation, ethylation or higher alkylation at the base level and also at the phosphate group level. -Alkylation takes place in the N or O position, with variable effectiveness depending on the base, position and MO used. -For example, **EMS** (ethylmethane sulphonate) is active in non-growth and with little letal, but mainly gives GC -> AT transitions; **MMS** (methylmethane sulphonate) is also active in non-growth and with little letal, but mainly gives GC -> TA transitions; **1,2-dibromoethane** is used in gaseous form (generally on Petri dish colonies); it gives GC -> AT transitions.
b.5) N-nitroso compounds:-These highly carcinogenic compounds induce point mutations as well as chromosomal breaks and alterations (acylnitrosamides and related compounds).**MNNG** causes methylations in 06 (guanine). They give very high mutation rates (up to 50% in *Escherichia coli*) with low mortality, but ëbeing highly toxic, they must be handled with great care.
b.6)Metals: -Manganese is known to cause transitions in bacteria; in yeast it causes mutations in mitochondrial DNA; -Hercury can cause chromosomal breaks in purine bases and mitotic accidents in eukaryotic organisms. Similarly, other metals (Co, Cr, Ni, Zn, Cu) can cause mutations.
b.7)Aflatoxins (mycotoxins):Aflatoxin B1 (AFB) gives GC - > TA transversions; it is inactivated by water and light, photosensitivity requiring handling under yellow light.
b.8)Hydrocarbons:-Whether aromatic or aliphatic, certain hydrocarbons have mutagenic and carcinogenic properties. -Benzopyrene is capable of producing transversions. It can bind covalently to DNA, causing torsion that can induce large excisions.

1.5.3- Other agents that disrupt the functioning of genetic material

a) Intercalating **agents - These** are products which are placed between the base plates of the double helix or which stabilise the shifts that occur following breaks. **Acridines** are intercalated between two purines and this can result in point mutations of the "frame-shift" type. Certain intercalating agents (acriflavine, ethidium bromide) are particularly active on extranuclear DNA and therefore induce cytoplasmic mutations.
b) Inhibitors of nucleic acid synthesis:-N-hydroxide, 2',3'-deoxyadenosine, 8-ethoxycafeine, 5-fluorodesoxypurine and other related compounds can cause chromosome breaks and interfere with DNA and RNA synthesis. **azaserine** inhibits the synthesis of purine bases and **urethane** that of pyrimidine bases. -It should be noted that these two products, in addition to their inhibitory role in the synthesis of nucleic acid precursors, can also act as alkylating agents.
c) Polyploidy inducers: There are mitotic poisons that disrupt spindle function: **colchicine, camphor, acenaphtene, phenylurethane, alkaloids** from *Vinca rosea,* **lead salts**, etc. are

capable of inducing polyploidy.

Other compounds may have the same consequences: plant growth substances (rndole, acëtonaphtalëne acid...), carcinogenic agents (benzopyrene, methylcholanthrene...).

NB:

Ames test for carcinogenic agents:The appearance of an abnormality is often linked to damage caused in the DNA. This test is used to estimate the carcinogenic potential of a substance.

Principle:-Test organism: Salmonella typhimurium (typhoid), His-. -ST does not multiply on synthetic medium containing Histidine. The test measures the reversion to prototrophy in the presence of mutagëne-S9 mix: preparation of rat liver enzyme (grinding, centrifugation). The supernatant containing the monooxygenase enzymes is kept at 80°c -This preparation is used to get as close as possible to the real iomme (the liver is the siëge of detoxification metabolic reactions) -GS: gelose overcoat used to spread the preparation. -C2 is twice the concentration C1 of the mutagenic product to be tested.

The test is positive if the number of mutants with the C1 product is at least double the number of spontaneous mutants.

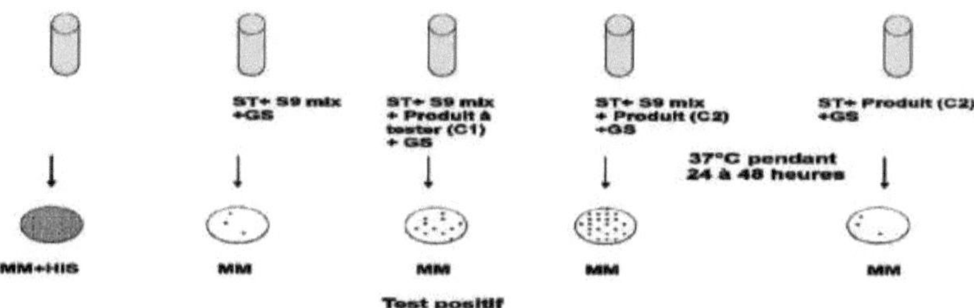

Fig 21: Illustration of the Ames test

1.6 Genetic heritage protection systems

1.6.1Protection against intrusion by foreign DNAs

In prokaryotic cells, there are systems for protecting the genetic heritage by destroying foreign DNA that may enter the cell and compete **with native DNA**.

The presence of such systems is much less obvious in eukaryotic organisms. **This "xenophobia"** calls on enzymes capable of destroying intruding DNA: **restriction endonucleases.**

These systems have been used extensively in a variety of bacteria, in particular *Escherichia coli*, and the native DNA is recognised as such and of course escapes lamination: this is due to mëthylation reactions on certain bases (adUnine and cytosine) which mark the DNA and enable it to be recognised at particular sequences.

The main source of mëthyl **is S-adenosylmethionine (SAM).** The dë system depends on the presence of several genes: in Escherichia coli, the hsdR, hsdM and hsdS genes code respectively for endonuclease, mëthylase and a site recognition factor.Recognition of native DNA applies not only to the bacterial gënophore, but also to the gënophore of the bacterium's specific bacteriophages (the phage DNA concerned is recognised and mUthylated, and is therefore protected against the action of the host's restriction endonuclease).

It should be noted that there are mUthylases other than those acting directly at the level of restriction endonuclUases:

-DNA adenine methylase (dam) carries out N6 mUthylation of adUnine in the GATC sequence, which is involved in postrUplicative repair. Depending on the case, endonucleases may or may not cut this modified sequence.

DNA-cytosine methylase (dcm) carries out a C5 mUthylation of the 2^e cytosine contained in the CC(A/T)GG sequence. Once again, the cut may or may not be made depending on the enzyme.

It should be pointed out that there are also DNA mUthylation reactions in eukaryotic organisms, whose role may be different: some are involved in controlling gene expression.

Bacteria contain 3 types of restriction enzyme, divided into two classes:

Class I consists of: Binary systems involving a separate endonuclease and methylase (type II). Their action requires the presence of Mg++, as in the case of the EcoRI system *in Escherichia coli*, whose site of action is a palindrome of 6 base pairs: the mëthylëes bases protecting the indigenous sëquences of this type are located at both ends.

There are many endonucleases of this type, which are widely used in genetic engineering: they are derived from different bacterial species and are capable of specifically recognising a large number of target sequences.

Enzymes of different origins with the same target sequence are referred to as **isochistomers.**

Class II consists of multimeric enzymes with both endonuclease and methylase activities (types I and III). These enzymes require the presence of ATP, Mg++ and SAM.

The EcoK or EcoB enzymes *of Escherichia coli* are examples of type I. The recognition site involves two sequences of 3 separate base pairs: it is responsible for binding the enzyme under methylation conditions identical to those in the previous case.

The cleavage takes place at a non-specific site located at a distance of 1000 base pairs from the recognition and binding site. The EcoPl or EcoP15 enzymes *from Escherichia coli* are examples of **type III**, which is quite similar to type I. Recognition and binding take place at an assymetric sequence of 5 to 7 base pairs, and cleavage takes place at a non-specific site fairly close to the recognition site (24 to 36 base pairs).

1.6- 2Protection against alterations affecting genetic material

- Protection of the genetic material is linked to the very structure of the cell:
- The cytoplasmic membrane and the nuclear membrane (in eukaryotic organisms) are permeability barriers for cations, nitrites and most molecules that can play such a role.)
- In eukaryotic cells, chromosome condensation, where it exists, protects their integrity at the time of segregation.
- The existence of a double hëlice, contribute to its stability: if one of the chains is altered, the genetic information is retained on the other.
- Enzymatic systems: catalase, peroxidases.
- Regulation of the physico-chemical properties of the cell environment (pH, osmotic pressure, etc.).

1.6.3- Repairing alterations

- Cells have repair mechanisms designed to rectify certain alterations to the genetic material: these are more or less specific enzymatic systems.
- Replication errors are normally repaired by the polymerases themselves, due to their exonuclease activity: proofreading repair occurs.
- In addition, there are post-replication control systems that correct any remaining faults (mismatch repair).

1.6.3.1) Repair by direct reversion of the lesion

a) Photo-restoration after ultraviolet (UV) mutation

- UV rays cause the appearance of thymine dimers (or more rarely other bases) which are repaired by the action of light.
- The dimer is cut by a photoreactivation enzyme, "photolyase" (or PR enzyme), which is activated by blue light (300 to 600 nm) while retaining the phosphodesoxyribose backbone.
- This system exists in bacteria (phr gene in Escherichia coli) but also in yeast.
- It should be noted that thymine dimers can also be repaired by other mechanisms (not specific to this lesion) which are not light-dependent (dark reactivation) and which involve excision or recombination processes (described below).

b) Desalkylation

- Alkylating agents are capable of transferring a methyl or more generally alkyl residue onto a base.
- Guanine is particularly sensitive (frequent formation of O6-methylguanine, or N7 methylguanine), but other targets are possible (O4 of thymine, sometimes N3 of purines, O2 of pyrimidines...).
- Escherichia coli has an inducible protein, alkylguanine-DNA alkyltransferase (product of the ada gene), which is capable of dealkylating alkylguanine without breaking the phosphodesoxyribose backbone.

1.6.3.2) Homology-dependent repair system

- Exploit antiparallel complementarity propriëtës to restore damaged DNA segments.
- In these systëmes, a segment of a DNA chain is eliminated and replaced by a newly synthesised nucleotide segment complementary to the model strand.
- One system (excision repair: e.g. BER and NER) repairs damage detected before replication. The other (post-replicative) system for repairing damage detected during or after replication.(MMR).

a) Excision repair

- These repairs are made possible by the double-stranded structure of DNA.
- The intact chain serves as a matrix for repairing the alternate chain.
- The most frequent repairs concern depurination sites or sites with thymine dimers, as these alterations are the most common spontaneous lesions.
- More generally, these repairs apply when the alterations lead to a change in the structure of the double helix (bulky lesions).

In Escherichia coli, there is a low amplitude excision system and a high amplitude excision system.

- The first, which is constitutive, affects around twenty nucleotides: it is the most frequent, occurring in 99% of cases, generally during spontaneous depurination.
- The second, which is inducible, can involve up to 10,000 nucleotides: however, the average number of excised nucleotides is around 1,500.

A.1)Base excision repair (BER)

- The stages of excisional repair are as follows (Fig 21):
- a)-Recognition of the site and cleavage of the N-osidic bond by DNA glycolyase with the appearance of an AP site; apurinic or apyrimidic)
- b)-Incision, i.e. breaking of the phosphodesoxyribose bridge at the diester bond by the action of an AP endonuclease (AP site; apurinic or apyrimidic).
- c) Excision of the nucleotides involved in the lesion and sometimes of nearby nucleotides

by the action of a deoxyribose phosphodiesterase.

- Excision involves a variable number of nucleotides.
- d) Restoration synthesis of the excised part by polymerases which use the intact chain as a matrix. In Escherichia coli, polymerase I appears to be responsible for synthesising short sequences and polymerase III for synthesising long sequences.
- e)-Soldering by a ligase, which in bacteria is NAD-dependent.

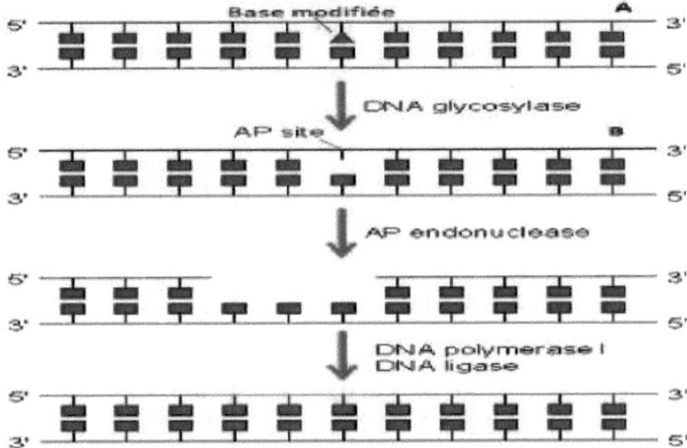

Fig 22: Base excision repair (BER)

A.2)Nucleotide excision repair (NER) system

- Rather than recognising a damaged particular base, the BER or NER system detects the deformation of the double helix caused by the pressure of an abnormal base.
- In *E. coli*, this complex is composed of three enzymatic activities encoded by the **UvrABC genes** (Fig 22), detects the deformation and cuts the damaged part at two sites on either side of the lesion.
- The UvrD exonuclease, excised (cut or eliminated) precisely 12 nucleotides: 8 nucleotides on one side of the damage and 4 on the other (in humans between 27 and 30 nucleotides are eliminated by a similar system).
- The 12-nucleotide gap is then closed by DNA polymerase I, using the template strand to produce an exact copy of the original DNA sequence.
- The DNA ligase then seals the new oligonucleotide.

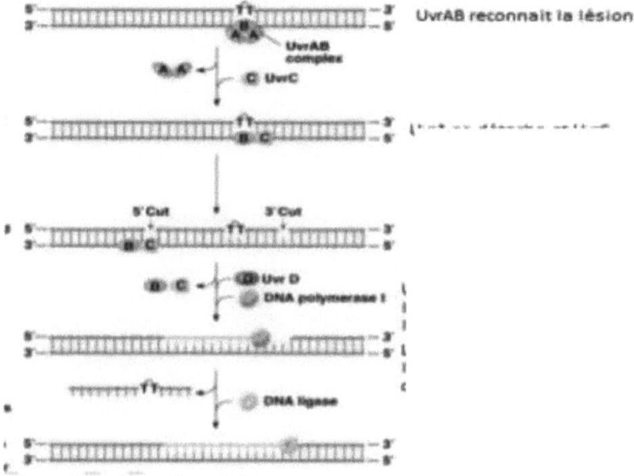

DNA polymerase fills the gap through complementarity
UvrA detaches and Uvrt clings to UvrB
DNA ligase Joins extrim itis
UvrD attaches to and iminates the oligonucleotide containing the lesion
Activiti endonuclease of UvrBC

Fig 23: Nucleotide excision repair (NER)

b) - Homology-dependent repair systems

b.1) MMR mismatch repair (post replicative)

- During replication, certain errors are recognised but cannot be repaired by the 3'-5' proofreading functions of the polymërase, as there is no modification of the bases or dëformations of the DNA, but it can consëquently generate mësappings between the bases.
- Another system called Mismatch repair (MMR) is then implemented, which will perform :
- 1. Recognition of mespaired base pairs
- 2. Determining the incorrect base.
- 3. Excision of the incorrect base and repair by neosynthesis of the excised fragment.
- It is a multienzyme repair system (Fig 23).
- **MutS** patrols the DNA looking for mismatches, and detects the strand carrying the error by distinguishing the degree of methylation of the two strands.

methyl)

- Helicase, MutU, breaks down the double helix by hydrolysing the hydrogen bonds.
- The MutH endonuclease incises and cuts the mutated strand at a certain distance on either side of the erroneous base.
- Pol III DNA resynthetises the missing fragment.
- The ligase reassembles the two pieces.

- The distance between the GATC site and the mësapair can be up to 1,000 base pairs long.

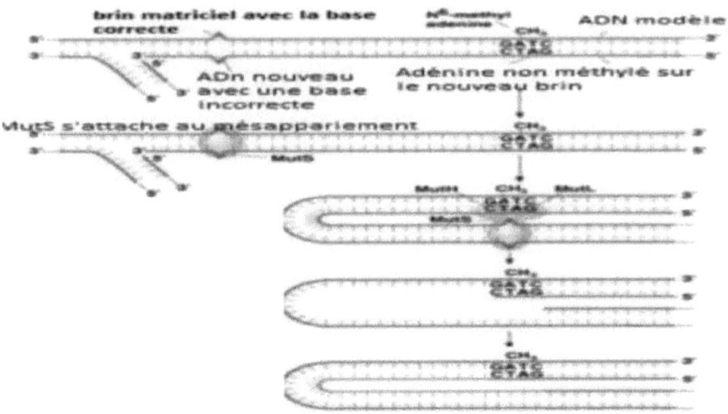

Fig 24: Repairing "MMR" mismatches

1.6.3.3- Repair of double strand breaks (DSB)

- Unlike the mechanisms described above, this mechanism allows mutations to be repaired simultaneously on both strands,
- We currently recognise two mechanisms: NHEJ (non-homologous end joining) and homologous recombination.

a) Repair via the NHEJ system (Non-Homologous End Joining)

- It is a non-conservative mechanism, meaning that it does not restore the initial DNA sequence, but only the continuity of DNA damaged by a double-strand break.
- This repair will thus lead to a change in the genetic information, generally a deletion, and therefore possibly to the appearance of a mutation for the gene concerned if the break occurs within a gene.
- This repair is made possible by the intervention of the dimeric protein **Ku70/80**, which interacts with the two DNA ends resulting from the break (Fig 24).
- This Ku protein has several activities:
- Nuclease activity: elimination of damaged nucleotides incompatible with end ligation. (In particular, Ku produces 5'-phosphate and 3'-OH ends compatible with strand suture by DNA ligase).
- Recruitment of other proteins involved in repair: a DNA-dependent protein kinase (DNA-PK), a specific DNA ligase (D) and a terminal transferase (TdT),
- Each time, the Ku70/80 protein checks whether hybridisation is possible or not, (2 to 4bases) the junction is stabilised and the break has been transformed into two single-stranded lesions.
- Any remaining single-stranded regions are then filled in by a DNA polymerase that uses the DNA strand just repaired by the Ku protein as a template.
- Ku also recruits a specific DNA ligase (DNA ligase IV in eukaryotes, DNA ligase D in bacteria) which sutures the strands together by reforming the cleaved phopshodiester bonds.

NB:

When the ends of the strand break are not clear, the Artemis protein acts by deleting a few bases. Its endonuclease activity is aided by the terminal transferase TdT and will enable the strands to be sutured together by the DNA ligase. It is during this so-called "end process" that certain bases can be destroyed, leading to mutations.

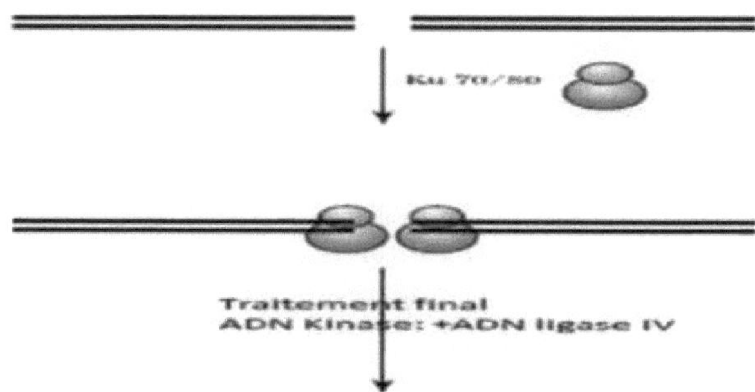

Fig 25: Repair using the NHEJ system

b) Repair by homologous recombination

This system allows repairs to be carried out in the absence of a matrix, or repairs to very extensive lesions.

- This is the mechanism that causes DNA molecules to exchange.
- Homologous recombination is initiated by a double-strand break in the DNA.
- The homologous recombination mechanism follows three main stages:

-A pre-synaptic phase

- 1)Resection Following the formation of a DNA double-strand break, homologous recombination is тШёс by the dëgradation, on either side of the break, of the DNA strand oriented in the 5'-3' direction.
- 2)Assembling the RAD51 filament
- Single-stranded DNA дёпёгё by resection is immëdiately bound by RPA, an abundant protein with high affinity for DNA.
- This protëine must be displaced to allow the Rad51/RAD51 recombinase (RecA in prokaryotes) to assemble.
- **A synaptic phase**, which involves searching for and exchanging a DNA strand with a similar (homologous) intact sequence in order to repair the break.

The mechanism behind this homology search is still relatively unknown.

- Once the homologous sëquence has been identified within the filament of RAD51, the RAD54(Rec A) protein slides along the filament and at the same time as it coils the "seeker" DNA strand and its complement to form a heteroduplex DNA, it detaches the RAD512 protein.
- The structure resulting from this strand ëchange, which comprises a heteroduplex double-stranded DNA and a displaced single-stranded DNA, is known as a D-loop (for Displacement loop).

A post-synaptic phase during which DNA strand exchange intermediates are resolved following the DNA synthesis required to restore the lost sequence at the break site.

l)DNA synthesis

- I . cxtreme 3' of the cassëe motecule finds itself hybridë to its intact homologue within the D-loop.
- A DNA polymërase ëtends this extreme by using the intact complementary DNA as a template. It is this svntliese step that allows the re-acquisition of the information lost at the

break site.

2)Resolution

• Once the synthesis step has been completed, the D-loop can be dismantled or resolved. Several sub-mechanisms branch off from this intermediate (SDSA (Svnthesis- Dependent Strand Annealing/DSBR (Double-Strand Break Repair)/BIR (Break Induced Replication).

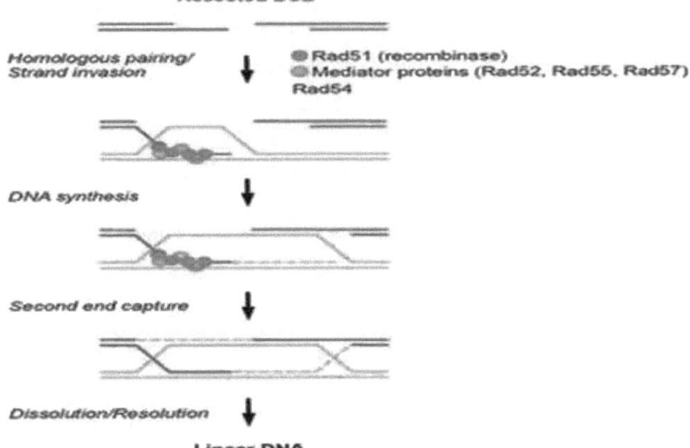

Fig 26 : Repair by homologous recombination

1.6.3.4- Reparation SOS (Save Our Selves) or by pass

• This inducible system is only active when the DNA has been damaged, and the presence of unrepaired DNA lesions induces emergency repair mechanisms: in the absence of these mechanisms, the cell's survival is at stake.

• The number of mechanisms involved in SOS repair varies according to bacterial species.

• In Bacillus subtilis, there are four different systems, compared with just one in *Escherichia coli.*

• The difference between these mechanisms is the nature of the proteins involved, which can mutate. The mechanism most often encountered is that demonstrated by E. coli.

A) Mechanism

• SOS repair will be triggered when the bacterium undergoes severe stress, resulting in DNA lesions that jeopardise its survival.

• Bacteria cannot stop replication (unlike eukaryotes) and must act more quickly to repair damaged DNA (Fig 26).

• The RecA protein binds to the DNA strands and initiates strand exchanges. It also has proteolytic activity, which cleaves the LexA protein.

• The LexA protein is a representative of the SOS system, and when it is cleaved, the genes involved in SOS repair are translated into large quanta.

• On the damaged DNA, the umuDC gene will be translated into proteins. This gene gives the bacteria mutagenic power and enables them to survive.

• Other genes in the SOS response namedës uvr will remove the damaged bases. They encode the endonuclease of the general excision system in order to repair the damaged DNA. The protëines UmuC and UmuD nëosynthëtisëes help polymërase III to overcome mutation barriers.

• The UmuD protein will automatically cleave into a modified UmuD' protein which will complex with UmuC. This forms an UmuD'2C complex that repairs the damaged DNA strand by allowing polymërase III to pass over the tesion barriers.

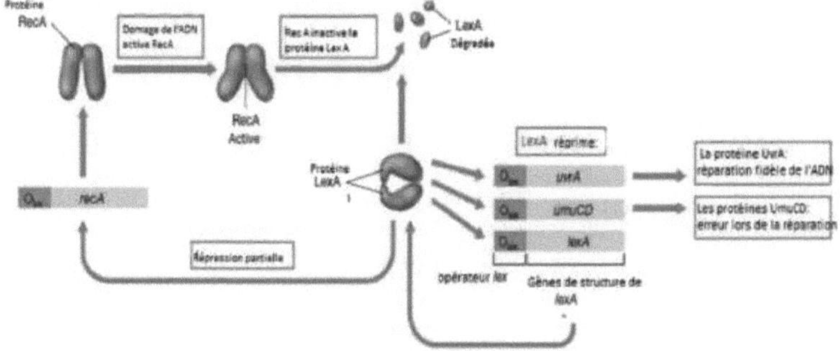

Fig 27: Rëparation SOS

NB:

<u>involvement of the SOS mechanism in bacterial revolution :</u>

1) Bacterial resistance to antibiotics

- The SOS response has been shown to regulate the expression of integrase from the class 1 integron of *Escherichia coli* and the VchIntIA intëgon of *Vibrio cholerae.*
- Integrons (genetic elements) are platforms on which gene cassettes are inserted and expressed. These cassettes may contain antibiotic resistance genes, among others.
- **2) Type of Gram**
- It is likely that the common ancestor of all Gram-positive bacteria is a Gram-negative bacterium that has lost its outer membrane and whose wall has thickened in response to the constraints of this new structure.
- This separation would have taken place 1.4 billion years ago
- A recent study has shown that mutations induced by SOS repair are responsible for the divergence between Gram-positive and Gram-negative bacteria.
- These differences between species are used by scientists to estimate when other organisms lived.

Chapter IV

Sexuality, parasexuality, cytoplasmic heredity and the mechanism of genetic recombination

l-Generalites

Recombination between gëtic matëriels from two different individuals occurs ташёге naturally through the processes of sexuality or parasexuality.

In fact, the notion of recombination goes far beyond this, and in general terms, genetic recombination is any process of exchange between DNA sequences.

General recombination is the most common phenomëne. It is essentially involved in sexual and parasexual processes in which information is exchanged between homologous sequences (or, at least, sequences with a high degree of homology). It results from interactions between whole homologous chromosomes or between chromosome fragments.

Recombination is necessarily intrachromosomal in prokaryotic organisms, whereas in eukaryotes there may also be interchromosomal recombination.

In eukaryotic organisms, recombination occurs frequently **during meiosis**, thanks to reductional division, which allows significant mixing of genetic information.

Recombinations are also possible outside sexual phenomënes, but they are much rarer and occur during mitosis of diploid cells (somatic recombinations, mitotic crossing-over: "meiotic like process", etc.) These recombinations are only easily detectable in heterozygous systëmes.

In prokaryotic organisms, recombination occurs during parasexual phenomënes and in viruses it occurs as part of the host cell. There are cases of recombination that do not require the confrontation of homologous DNA sequences: these are known as site-specific recombinations. They involve viruses, internal transposable elements and sometimes plasmids.

Sexuality or parasexuality results in a redistribution of al^les within a population; this can confer a selective advantage on cells, enabling them to cope with changes in environmental conditions.

2-Sexuality in eukaryotes

Sexuality allows the mixing of genetic information from two cells, through specific and relatively complex processes: there is a confrontation between two complete gënotypes.

Typical sexual phënomënes are only found in certain eukaryotes. When a biotype of a normally sexual species has no sexuality, it is called an imperfect form.

Sexual reproduction involves several phases: first there is confrontation of the genetic material of two individuals, then possibly recombination followed by the segregation of new genotypes. The mechanisms involved in the sexuality of microbial eukaryotes are extremely varied and some are highly complex.

2.1 Notion of biocycle

The notion of sexuality is linked to that of the biocycle, which alternates between the haploid phase (n chromosomes) and the diploid phase (2n chromosomes). During these two phases, cells can divide through the mitotic process, resulting in the growth or "vegetative" multiplication of individuals.

Depending on the species, this vegetative growth or multiplication phase affects haploid cells alone (haplobiontic cycle), diploid cells alone (diplobiontic cycle) or both (haplo-diplobiontic cycle).
The two main sexual phenomena that link the haploid and diploid phases are cell fusion (copulation, hybridisation, conjugation...) and meiosis. During fusion, two haploid cells copulate to give rise to the first cell, the diploid phase or zygote. During meiosis, a diploid cell divides and gives rise to the first cells of the haploid phase or spores (in the strict sense of the term).
These two phenomena can follow each other in either direction with a minimum of intermediates (sometimes none at all): this explains the existence of haplobiontic (meiosis follows fusion) or diplobiontic (fusion follows meiosis) sexual cycles.

2.2 The mechanism of sexual fusion

According to **Esser and Kuenen**, different types of sexuality can be distinguished in eukaryotic microbial organisms. These different possibilities are found in moulds.

- Presence of defined sexual organs (male or female) :

a) on the same individual: the lines are hermaphroditic and any two lines may cross (for example: various Euascales and Uredinales).
b) on different individuals: the lines are male or female and the cross can only be made between males and females (this case is exceptional and concerns rare species).
Ascomycetes and Phycomycetes, for example Achlya).

- bX^No characteristic "sexual organ".

a)an individual can be a donor or an acceptor in a cross: the lines in a situation close to hermaphroditism but there may be incompatibilities between lines preventing any lines from crossing (this situation is common for many moulds).
b) an individual is of a given sign (equivalent to a given sex) and can only cross with an individual of the opposite sign (for example: various yeasts, Mucor...).
When fusion affects morphologically identical cells, there is **isogamy**, and when it affects different cells, there is **anisogamy or heterogamy** (sometimes referred to as male and female cells). When fusion is possible from individuals belonging to the same clone (of single-cell origin) or from cells of the same individual or individuals of the same clone, there is **homothallism.**
If fusion requires individuals or cells from individuals of different clones, there is **heterothallism**: in the latter case, the different clones have a clearly defined sexual type (a or a, + or -...) and fusion can only take place, in principle, between two clones of different and complementary types.
Most of the time, it only has two sexual types: this is found in many yeasts (Saccharomyces cerevisiae...), moulds (Mucor...), algae (Chlamydomonas...) and protozoa (*Paramecium aurelia*...).However, there are more complex cases: the yeast Schizosaccharomyces pombe can be both homo- and heterothallic, it has three sexual types due to three different alleles h^+, h^-. and h90.
In yeast, homo- and heterothallism are controlled by transposition phenomena; cell fusion can be divided into two stages: plasmogamy or cytoplasmic fusion and karyogamy or nuclear fusion.
Plasmogamy can be facilitated by substances released by the cells. For example, in the yeast *Saccharomyces cerevisiae*, the a and a strains release peptide substances called pheromones: factor a (composed of 11 amino acids) and factor a (composed of 13 amino acids) (encoded

by the MFa and MFa genes). a and a factors bind to cell receptors of opposite sign and block cells in Gl phase, which facilitates plasmogamy. Plasmogamy may be absent: in this case, there is only an exchange of nuclei between two cells, and this exchange is generally reciprocal. This is the case, for example, *in Paramecium aurelia.*

2.3 Mechanism of meiosis

Meiosis is a succession of two divisions: the first is reductional, because it modifies the ploi'die, and the second is equational, without modifying the ploi'die (Fig 27) . There is only one duplication of chromosomes during this process.

2.3.1 First division phase

a)Prophase

This phase is complex and is traditionally divided into several stages that are not visible or only slightly visible to eukaryotic microbial cells due to the absence of chromosome thickening:

1-Individualisation of the chromosomes, which unlike during mitosis are not yet cleaved into chromatids (**leptotene stage**).

2-Pairing of homologous chromosomes (**zygote stage**). Pairs of homologous chromosomes are called "bivalents".

3-Very reduced or non-existent thickening phase, unlike in higher eukaryotic organisms (**pachytene stage**).

4-Chromosomes split into chromatids, resulting in DNA duplication (**diplotene stage**). Chromatids from the same chromosome are known as error chromatids. At this stage, **chiasmas** may appear, i.e. areas of contact between chromatids where crossing **over** may occur.

b) Metaphase, anaphase, telophase

In microbial eukaryotic organisms, metaphase and anaphase are not always clearly distinguishable (even electron microscopy can only show the evolution of the spindle and the plates that separate and orientate, marking the first division). As with mitosis, the nuclear membrane does not disappear in many families (particularly in yeast).

There is no interphase between the two divisions of meiosis.

2.3.2 -Second division stages

The second division, or equational division, resembles normal mitosis: the exanen of the plates and spindles makes it possible to follow its progress, and in the end 4 groups of n chromosomes are formed. When the nuclear membrane does not disappear, the nucleus remains cohesive but gradually undergoes deformation before dividing; this division is followed by that of the cytoplasm with the appearance of the first haploid cells.

In some species, one or sometimes two additional mitoses may occur during the sporulation process, bringing the number of spores to 8 or 16.

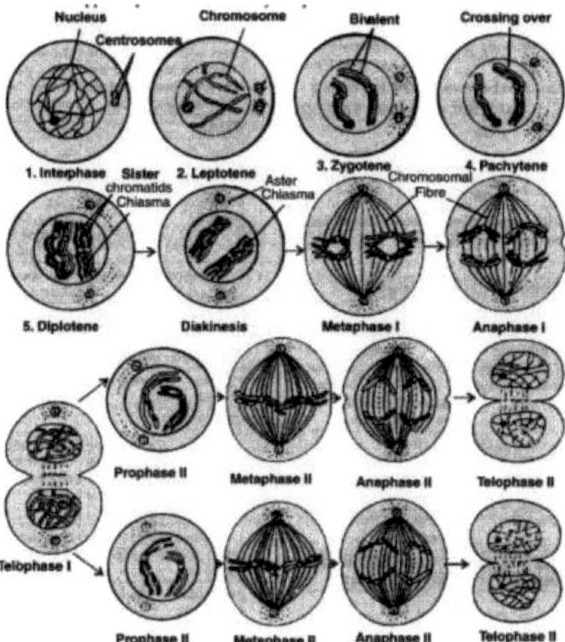

Fig 28 : The different stages of meiosis

3-Parasexuality

Para-sexuality is the normal mode of genetic mixing in prokaryotes, but parasexual phenomena are also found in certain eukaryotes. pansexuality can occur through hybridisation (somatic fusion) of two cells, in which case two complete genomes are brought together and a diploid cell is formed if there is karyogarny. This occurs in yeasts and moulds.

Parasexuality can also occur through the intervention of a receptor cell (whose genetic material constitutes the endogenous) and genetic material from another (exogenous).

The exogenote can enter in various ways:

a) Directly, i.e. without the intervention of a vector: either with obligatory cell contact (conjugation), or without cell contact (transformation).

b) By means of a vector of cellular origin, generally a plasmid (transformation or conjugation by plasmid, sexduction) or by means of an external vector, generally a virus (transduction).after entry, the exogenote can have several fates: either it is degraded by the host's restriction systems, or it remains in a stable exogenous state, or it is involved in a recombination process.

3.1Recombination in viruses

It occurs during a mixed infection. Genetic mixing takes place between the two genophores within the host cell, usually during the replication process. There may be 4 to 5 recombination operations during a single lytic cycle: the mechanism is that of simple or reciprocal crossover. In even T bacteriophages (T4), there is a heterozygosity phenomenon corresponding to the presence of 2 strands carrying a different allele, resulting in the formation of a partial heteroduplex.

During a mixed infection, the genophore of a virus may be encapsidated in a hybrid capsid composed of proteins from the 2 "parents". In this case, there is no genetic mixing as such and the hybrid obtained is not a recombinant. This phenomenon is known as "phenotypic mixing". The most extreme case is when the genophore of one phage is found in the capsid of another: this is called transcapsidation or genotypic camouflage.

3.2 Bacterial parasexualtte

a) Transformation

The transforming principle is that DNA enters the cell and is then recombined with endogenous DNA: the introduced trait is not added to the existing traits but takes the place of one of them and becomes hereditary.

Competence is an induced characteristic due to a surface modification controlled by a protëic factor: it depends on the plysiological state and environmental conditions.

In Streptococcus pneumoniae (Pneumococcus), and group H Streptococcus in general, the natural transformation process involves the following ëtages (Fig 28):

1-DNA in a bicatëary state binds to a receptor site (30 to 80 sites per cell).

2-Pënëtration into the cell involves a membrane endonuclease which acts as a DNA translocase by dëgrading one of the strands and promoting pënëtration of the other.

3-Integration occurs by recombination with displacement of the recipient's homologous chain to form a heteroduplex segment.

Recombination is only possible when there is a high degree of homology between endogënote and exogënote.The various exogënote markers have variable transformation efficiencies (VHE: very high efficiency, HE: high efficiency, IE: intermediate efficiency, LE: low efficiency) depending on their sensitivity to repair systems (repair systems tend to "erase" homology differences, and therefore reduce transformation efficiency).Transformation involves DNA fragments smaller than 40 kb. It requires cellular competence, which occurs at the end of growth. Intergeneric transformation is also possible, particularly between certain Streptococcus and Staphylococcus bacteria.

The mechanism of transformation is markedly different in Gram- bacteria because of the peculiarity of their wall, as has been studied in *Haemophilus influenzpe.*For this species, competence is linked to an increase in the level of Lipopolysaccharides in the outer membrane, with a qualitative modification: no competence factor could be isolated. Moreover, it is not certain that donor DNA entry occurs in single-stranded form, it seems that Гopëraňon double-to-single-stranded passage occurs inside the cell.pënëtration requires recognition of a 10 bp sëquence (11 bp in Hemophilus).

Cellular competence depends on the presence **of pili, and** transformations can also occur by transfer of non-conjugative plasmids. In certain specific cases, these plasmids can play the same role as the F factor in sexduction.

NB: Artificial transformations can be achieved by using well-chosen plasmids. In general, during these transformations, the pënëtration mechanism is slightly different: there is no single-chain onversion. Moreover, in these cases, the efficiency of the plasmids used (which are generally artificial plasmids) is superior to that of DNA of chromosomal origin.

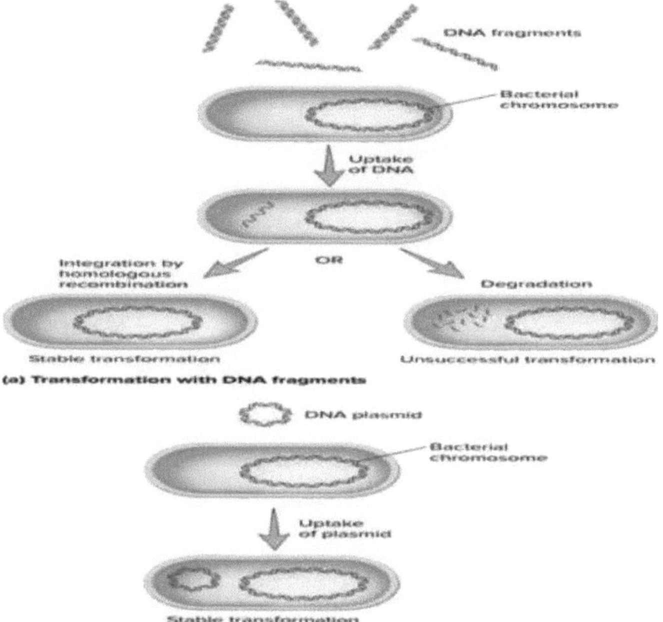

Fig 29: Bacterial transformation

b) Transfection

Transfection is a phenomenon similar to transformation, which involves phage DNA and occurs in many bacterial species: *Bacillus subtilis, Haemophilus influenzae, Staphylococcus aureus, Escherichia coli, Salmonella typhimurium, etc.*

In many cases, this results in the cell producing viruses. As with transformation, there are competence genes which, in Gram- bacteria (*Escherichia coli, Salmonella typhimurium*), may be linked to a deficiency of lipopolysaccharides in the outer membrane.

For these strains, the efficiency of transfection depends on the calcium concentration of the medium. Uptake of the viral material by a competent cell is not always effective: in fact, good integration requires a certain degree of homology between endogenous and exogenous DNA, and in this case, homology is often insufficient. Furthermore, as the integrated DNA is in single-chain form, it cannot generally replicate.

It has also been observed that transfection efficiency is greater for receptor strains deficient in their recombination system (rec-), but is increased if the receptor strain is lysogenised by another bacteriophage.For example, *Bacillus subtilis* lysogenised by phage ϕ105 is highly efficiently transfected with DNA from phage ϕ SP02 and *Staphylococcus aureus* lysogenised by phage PI1 is a more efficient receptor.

c) Transduction

Transduction is gene transfer by means of a virus (bacteriophage) which is the vector of the exogenote. This phenomenon is fairly common in bacteria (Enterobacteria, *Bacillus subtilis*, Pseudomonas, Staphylococcus, Streptococcus, etc.).

There are two types of transduction:

c.1) Restricted or specialised translation

It uses tempered bacteriophages (Fig. 29) which are capable of attaching themselves to the chromosome of the recipient bacterium as a prophage and multiplying with it (lysogenic systems), the classic example being *Escherichia coli* and the bacteriophage lamba(^) (Fig. 30).

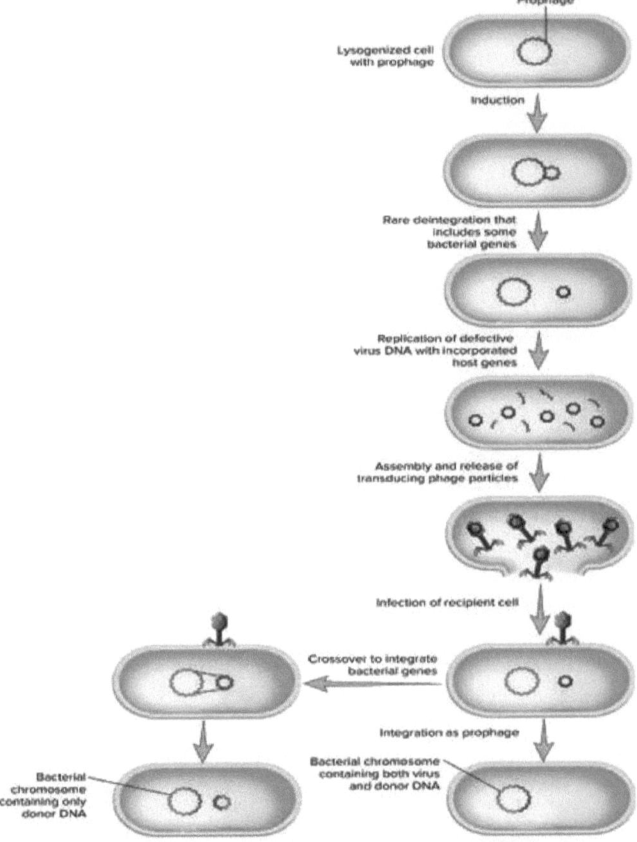

Fig 30 :Specialised transduction by a temperate bacteriophage

This phage binds as a prophage to a specific point on the chromosome, close to the gal locus. When the prophage transforms into a virulent bacteriophage, some of the bacteriophage forms that carry bacterial genes may have lost one or more of their own genes: they are said to be defective (^d). Such a bacteriophage carrying the gal bacterial gene is called ^dgal.

Bacteriophages carrying the exogenote can transmit it to the bacteria they infect (this transfer is rare, its frequency is 10-5; low frequency transduction).

Bacteriophages like ^dgal cannot replicate or transduce at high frequency on their own: they require the presence of normal bacteriophages known as helper bacteriophages.

In rare cases, it may happen that the exogenote does not integrate (it functions but does not replicate), in which case there **is abortive transduction.**

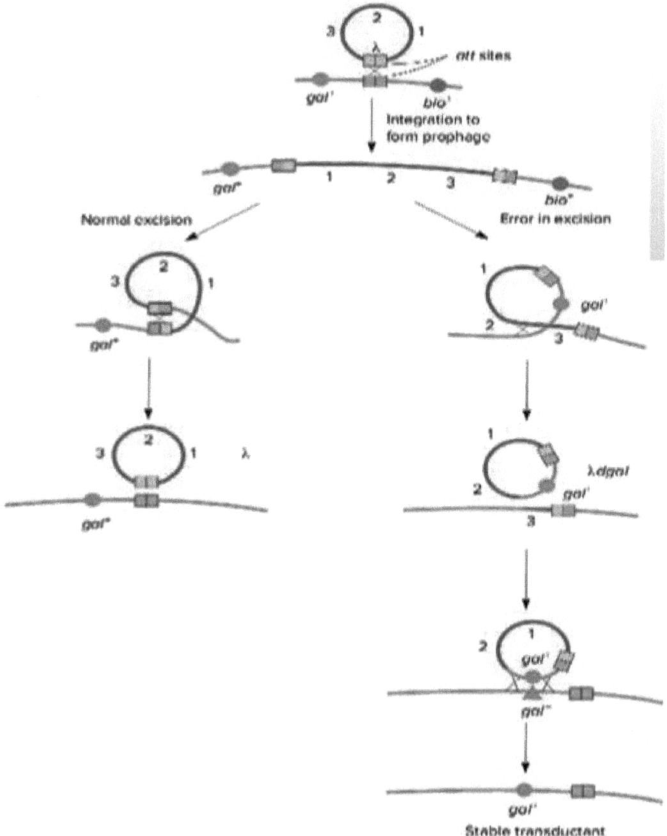

Fig 31 :Lamba phage transduction mechanism in *E. coli*

c.2)General transduction

There are cases of transduction that allow the transfer of any genes. These transductions can take place by several mechanisms.

c.2-1) Bacteriophage *P22/Salmonella typhimurium* pair:This transduction takes place at low frequency (10^{-3}). At the time of virus formation which precedes lysis, the bacteriophages can integrate fragments of the bacterial chromosome previously destroyed by the viral DNAase in association with their own genetic material, or in its place (Fig.31).

Unlike the previous case, where transduction can only occur through induction of a lysogenic bacterium, this case of generalised transduction can occur during a normal lytic cycle. Abortive transduction is common in this type of generalised transduction: the virus cannot integrate into the host chromosome if its genome is too incomplete, or even non-existent.

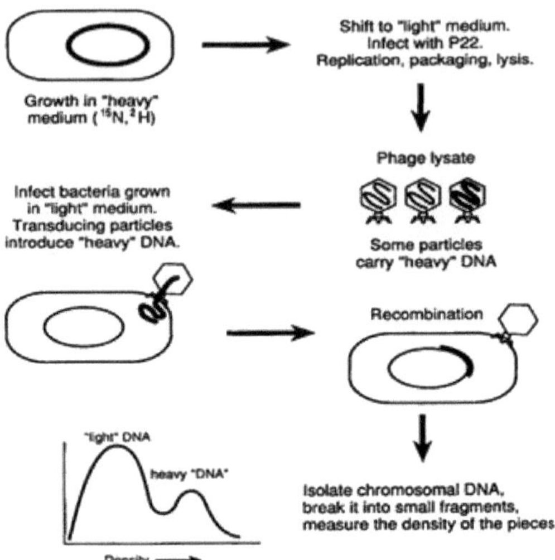

Fig 32: Physical transfer of DNA in generalized transduction (Example of phage 22)

It should be noted that phage P22, although similar to other phages in the ^ group (^,ф80...), has a different mechanism for controlling the lysogenic and lytic cycles: inactivation of the lytic cycle repressor is not RecA-dependent but results from synthesis of a phage antirepressor. Under lysogenic conditions, the antirepressor is not synthesised.

Staphylococcus aureus has several P22-type transducing temperate phages: ф 11 (with a permuted circular DNA of 45 kb), ф 12, ф 13. Similarly, phages SP10 from *B. subtilis* and CPT1 from *Vibrio cholerae El Tor* belong to the same category.

c.2-2)A second type of generalized transduction involves a lysogenic bacterium and a temperate bacteriophage which is able to integrate at any point on the chromosome and not at a specific site (for example the Liu bacteriophage).

During transduction with gu, the capsid can contain various combinations of the phage genophore and various fragments of the host genophore or plasmid.

c.3-3)A third type uses phages that remain in the extrachromosomal state after infection *i*.e. the PI phage *from Escherichia coli*. The prophage behaves like a monocopy plasmid. The mechanism of this type of transduction is poorly understood.

d) Lysogenic conversion

Following a lysogenic state, a bacterium can acquire new properties that are apparently unrelated to this state. This is known as lysogenic conversion. In fact, the modification is due to the presence of phagic genes that cause the expression of the property.

This phenomenon occurs in *Salmonella anatum* and *Salmonella holera- suis*(*enterica*): there is a modification of the surface polysaccharide, or O antigen, following lysogenisation by bacteriophage e or other bacteriophages.

Similarly, the production of erythrogenic toxin by Lancefield group A Streptococcus and Clostridium botulinum types C and D toxins is linked to lysogenisation.

This mechanism explains the interspecific conversions described in Clostridium: non-lysogenic type C Clostridium botulinum can be converted into Clostridium novyi by the bacteriophage specific to the lysogenisation of this strain.

e) Conjugation

This phënomëne occurs in a large number of bacterial species (Entërobactëries, Rhizobium, Streptococcus, Streptomyces...). It has been particularly well studied in *Escherichia coli.*

e.1) Conjugative plasmids

Conjugation involves contact between two cells: a specific pairing takes place between donor and recipient cells. The donor cell has parietal structures involved in the pairing and transfer of genetic material: **the sexual pili**. These pili belong to 2 categories, F and I, which can be distinguished by phage typing (specific RNA bacteriophages) and immunology.

After contact, pili are likely to retract, causing the cells to stick together. The formation of pili is linked to the presence of conjugative plasmids in the donor cell.

In principle, the receptor cells must be of the "female" type, i.e. they must not themselves possess conjugative plasmids. Cells possessing a class of plasmids are poor receptors for this plasmid or neighbouring plasmids (remember that there are incompatibility phenomena between the different classes of plasmids).

In *Streptococcus faecalis*, peptides involved in the conjugative transfer of plasmids have been identified, and are known as sex hormones or pheromones. They promote contact between donor and recipient cells.

Receptor cells do not contain a plasmid and have in their genome a determinant (BS) for a surfactant and one or more determinants (cA, cB...) for a pheromone.

Pheromones are thermostable, protease-sensitive hydrophobic octapeptides that induce the production of a BS-complementary AS surfactant in donor cells.

The donor cells have the same determinants at the genophore level, but the plasmid they contain carries a specific receptor for a pheromone (RcA for pheromone cA), the product of which activates the AS determinant.

The plasmid also carries a determinant (IcA) whose product inhibits pheromone autoproduction.

In many cases, for example with the pAMβ1 plasmid, we have a binary system: the phëromones cADl and cPDl are excreted by the receptor bacteria, while the inhibitory peptides iADl and iPD1 are excreted by the donor strains.

Plasmid transfer can result in the recipient cell acquiring new properties from the transmitted genes, such as resistance to an antibiotic (this poses a real problem in the case of pathogenic recipient strains).

It should be noted that plasmids carrying resistance (factor R) are classified into 2 categories, which use pili I or pili F for their transfer: pili I are induced by the colicinogen factor ColEl (pColEl), whereas pili F are induced by factor F: both types bind various phages.

e.2)F factor

Escherichia coli factor F (Fig 32), or fertility factor, is an **episome-type** conjugative plasmid, i.e. capable of integrating with the chromosome. It is made of DNA and has a volume corresponding to 1/50th of the bacterial chromosome: its molecular weight is 62.10^6 Da and it contains 94,500 base pairs.

Factor F has genes for autonomous replication, for the formation of sex pili, for conjugative transfer and contains insertion sequences (2 of IS3 type and 1 of IS2 type).

Cells which possess the F factor are called F+: these are the donor cells or "male" cells. The number of F factors per cell is low: 1 to 2.

There is stringent control of the number of copies. This phenomenon can be explained by positive regulation involving attachment to the cytoplasmic membrane and therefore coupling with growth and division, or more probably by negative regulation linked to the presence of a repressor product encoded by the plasmid.

The receptor cells, or "female" cells, are known as F- cells. During cell-to-cell contact via the pili, the F- cell recoils factor F and becomes F+.

The mechanism of transfer has been elucidated: an endonuclease, encoded by the traY and Z plasmid genes, cuts a strand of the plasmid at the oriT sequence and the 5' end thus formed migrates towards the recipient cell while the transfer is taking place.

replication by the "rolling circle" mechanism.

Thanks to this mechanism, the donor cell remains F+ even though it originally possessed only one F factor. Contrary to what was thought, it does not appear that the DNA passes via a pili, but rather through pores and specific bridges (involving the gene (traD) with the possible intervention of a pilot protëin.

However, the presence of pili is necessary for the expression of factor F, so F+ mutants without pili have an F- phenotype.

The 5' end of the strand that enters the recipient cell binds near the entry pore and remains attached until the replication process is complete. The TraT protein (product of the traT gene) exerts a negative control on transfer by binding to the inner membrane and blocking the sites involved in conjugation.

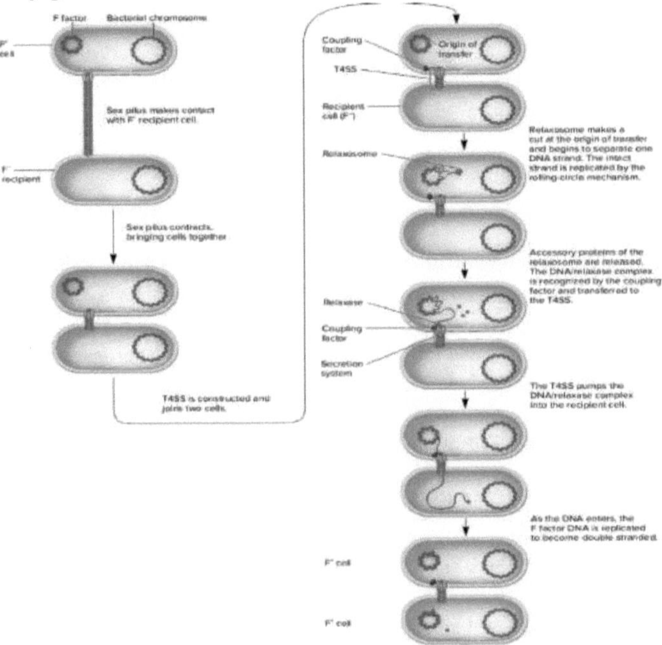

Fig 33:Conjugation due to the F factor

e.3) Chromosome transfer

During a conjugative phënomëne. and parallel a Гёсьапдё of F plasmids which takes place for a large number of cells, a much rarer (10-4 frequency) chromomic gënëtic material ëchange (low-frequency chromosomal -ic conjugation) can occur.

The F+ cell acts as a donor and the female cell as a recipient: the donor transmits only part of its genetic material (exogënote). Gënëtic mixing is only effective after a process of gënëtic recombination between this exogënote and the endogënote of the recipient cell.
There are mutants that give a high recombination frequency. These are called **Hfr mutants**: they are derived from F+ cells.
In Hfr mutants, factor F integrates into the chromosome by a process similar to that of bad^riophage ^ integration (Fig 33). During cell division, the DNA of an Hfr mutant replicates normally. When there is contact with an F- cell, a rolling circle replication occurs and only one strand of the chromosome is transfected, the other remaining in the cell, which explains why it retains the Hfr propriëtë.
In addition, the position of a gëne on the chromosome determines the frequency of recombination. It should be noted that in the presence of a large number of Hfr cells, the F-receptor cells have a high death rate: this is known as "tetal zygosis".
Conjugation can sometimes affect cells from different espëces, for example *Escherichia coli* and Salmonella, Serratia or Shigella (interspëcific conjugation).
In *Vibrio cholerae*, there is a 68 kb conjugative plasmid or fertile factor P: there are also P+ and P- strains. However, mobilisation of the chromosome is very rare because there is no insertion sëquence on the plasmid.

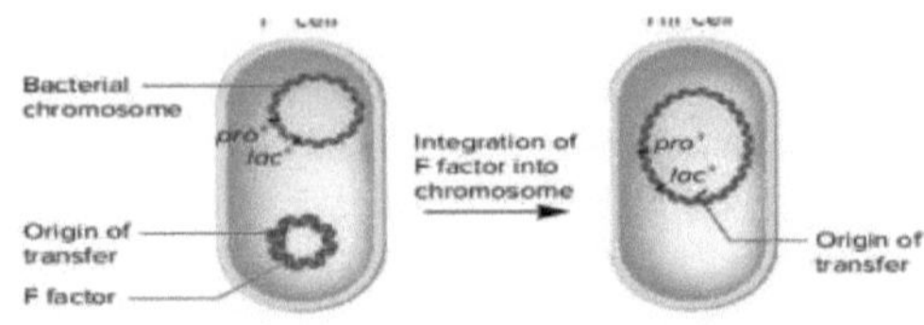

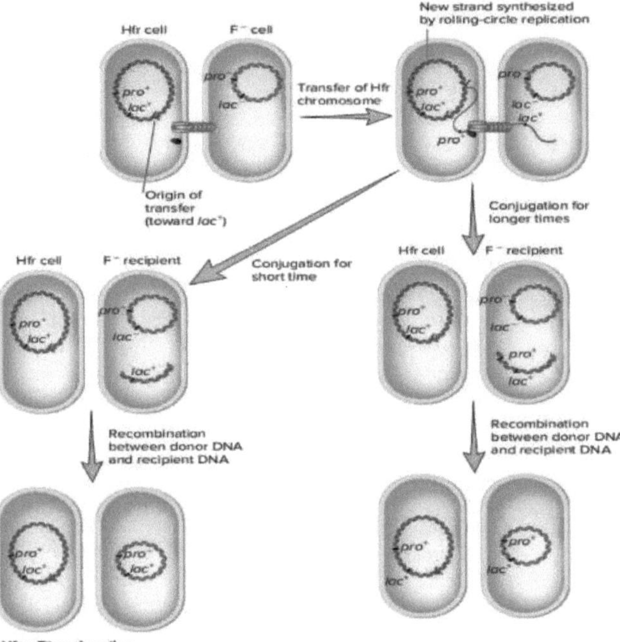

Fig 34 :Creation of Hfr strains and Hfr x F- conjugation

e.4)Sexduction

The Hfr trait can become F+ again by mutation. Under certain (rare) conditions, factor F can take along a fragment of the chromosome which was adjacent to the binding site: it is then called factor F'. As it can be easily transferred back to F- cells, it can transfer this chromosome fragment at the same time.

Depending on the mechanism of formation, which may involve various recombinations, type I F' is formed (where the F genome is incomplete) or type II (where it is complete). This phenomenon, known as sexduction, is quite similar to the transduction by viruses (Fig. 34).

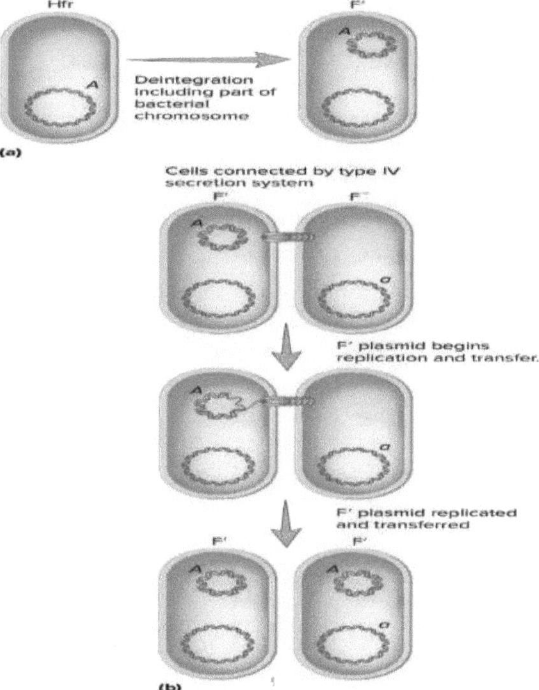

Fig 35:The F' conjugation

It is possible to construct F or F' factors containing an insertion such as the transposon TnlOLthe homology thus obtained with the gënophore (at the level of Tn 10) facilitates its mobilisation. This is used in gënëtic genetics.

4)Somatic fusion of eukaryotes

It is common in fungi (moulds) without a sexual cycle or with rare sexuality. After cell fusion, genetic mixing occurs by mitotic recombination.

The various stages are as follows:

- Formation of heterokaryons by fusion (anastomosis) of two haploid mycelia.
- Rare nuclear fusion (frequency 10^{-6}). The haploid and diploid nuclei are found mixed together in the mycelium and eventually in the conidia.
- Mitotic cross-over (frequency 10^{-2} to 10^{-3}), this is an intrachromosomal recombination occurring at the 4-strand stage and affecting only 2 strands. The recessive markers present in the heterozygote remain masked.
- Haploidisation (frequency 10^{-3}). Haploidisation can occur directly in non-recombined diploids causing interchromosomal recombination or after mitotic crossing-over, which corresponds to a true parasexual event reflecting intra-chromosomal recombination.

5)General recombination mechanism

General recombination occurs during sexual and parasexual processes when two haploid genomes (one of which may be partial) come into contact. It takes place between homologous chromosomes or chromosome fragments.

Recombination is said to be reciprocal when there is a balanced exchange of markers between the two confronting elements; it is said to be non-reciprocal when the exchange is not balanced (conversion phenomenon) or when the exchange is not mutual (case of the integration of a "donor" fragment on a "receptor").

Among the different types of recombination, a distinction is made between interchromosomal recombination, which leads to a reciprocal distribution of characteristics, and intrachromosomal recombination, which takes place via a crossing-over.

Intrachromosomal recombination may be inter- or intragenic and may be reciprocal (most common in eukaryotic organisms) or non-reciprocal (common in prokaryotic organisms, rarer in eukaryotes).

5.1 Interchromosomal recombination

It consists of a new distribution of homologous chromosomes at the time of mëiosis (Fig. 35).naturally, this mechanism only occurs in organisms with several chromosomes, i.e. eukaryotic organisms, usually in a sexual process, more rarely parasexual.

Each daughter nucleus recoils one chromosome from each bivalent during sëgrëgation and the repartition is gënëral at random. As a result, parental combinations are rare and recombinants predominate. This is all the more true when the number of chromosomes is high.

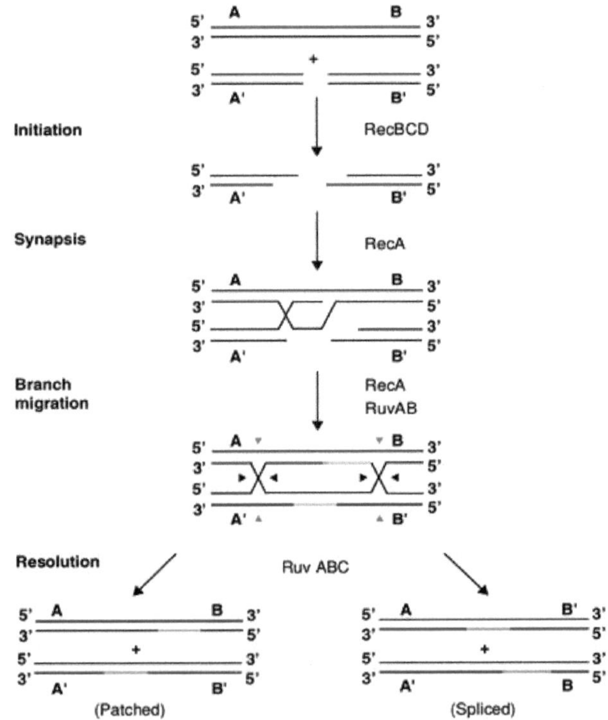

Fig 36:Stages of homogeous recombination in *E.coli*

5.2 -Intrachromosomal recombination.

a) <u>Recombination by pairing two single strands:</u>This mechanism has been successfully ëlë ëШ&ë thanks to the T4 ba^riophage. This is a non-reciprocal recombination involving the following ëtages:

- Action of an endonuctease which splits one strand of DNA from each duplex (single-chain break).
- Action of an exonuctease which ëк^к the break zone. At this level, stabilisation takes place using a protëic factor (SSB protëine in *Escherichia coli*).
- Pairing of intact monocatëary regions, resulting in the appearance of a hybrid bicatëary DNA region.
- Extension of the hybrid zone then new breaks to ë eliminate unnecessary DNA fragments.

b)Recombination by single strand displacement:This type of recombination appears to occur occasionally in prokaryotic organisms and to be the rule in eukaryotes. In most cases, it is a reciprocal system.

Depending on the modële, one or 2 strands are replaced.

The Holliday modële involves the displacement of two strands and comprises the following steps (Fig 36)

- Endonuclease cleavage of one strand of each duplex
- Dë replacement of free strands
- Rë union of these strands: the classic crossover pattern is formed
- Dëplacement of the crossing-over with ëtrical exchange of the chains and formation of two hëtëroduplexes whose size increases (branch migration)
- 180° rotation of the moiM of the formëe structure with dëcroisement of the crossed strands (isomërisation). The figures obtained are known as Holliday figures.
- Cutting of one strand of each duplex, which can take place in two planes. In both cases, a hëtëroduplex is formed, but only one of the two leads to recombination.

A slightly different mechanism, which takes account of DNA neosynthesis, can be envisaged (**Meselson-Radding model**): it involves only the displacement and capture of a single single strand. Its stages are as follows

- Cutting of a chain by an endonuclease.
- Displacement catalysed by DNA polymerase or helicase (found in Escherichia coli), with HDP and SSB proteins protecting the displaced chain.
- Neosynthesis of DNA on the non-displaced end of the broken chain.
- Incorporation of the displaced DNA into the receptor double chain thanks to the intervention of protein factors (RecA in *Escherichia coll*) and, possibly, gyrase, with the formation of a loop (D-loop): a crossover is initiated.
- The loop is broken by a nuclease and the displaced chain is assimilated by pairing with one of the receptor chains, while the other is eliminated.
- Isomerisation of the structure, known as the cross-strand exchange structure, using 180° rotations (Holliday figures), which effectively concludes the crossover.
- Possible displacement of the crossover with symmetrical exchange of chains and formation of two heteroduple x of increasing size (branch migration).
- Crossover cleavage by nucleases and repair by polymerases and ligases.

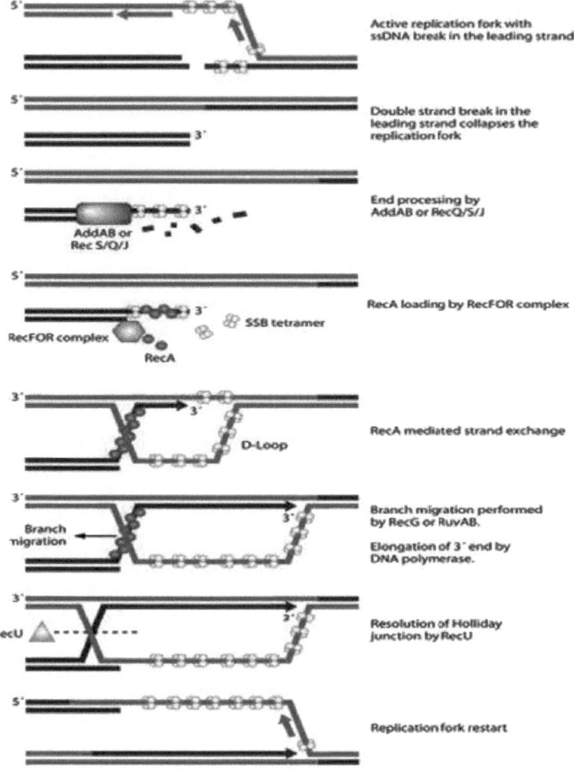

Fig 37:Recombination in *B.subtilis*

5.3 Intragenic recombination

Intragenic recombination is a recombination that occurs within a gene involving two alleles. Intragenic recombinations have made it possible to interpret allogeneic transformation phenomena in Pneumococcus:- a normal strain of Pneumococcus is sometimes obtained by transforming a mutated strain using DNA from a different mutated strain (mutations in the same gene). The presence of intragenic recombination also makes it possible to interpret conversion phenomena.

5.4 Mechanism of recombination between non-homologous chromosomes or chromosome fragments

These recombinations do not take place randomly, like general recombinations, but at specific sites, depending on whether one or both of the elements involved have one or more specific sites.

In cases of double specificity, there are sequences with homology, but their size is very limited: the homology concerns ten or fewer bases.

These recombinations are referred to as site-specific recombinations and sometimes as 'illegitimate recombinations'. When these recombinations occur between two sites on a single DNA molecule (which may be a hybrid molecule), they can lead to the inversion or deletion of the sequence between the two sites.

a) Recombination of sites with dual specificity: the case of the integration of specialised transduction bacteriophages (tvpeA):Recombination takes place using complementary specific sites, one of phage origin and the other of bacterial origin. The best-known example is that of **bacteriophage A**: the phage site is called_attP or (BOP') and the bacterial site attB or POB'. The bacterial site is located between the gal and bio genes.

Hybrid sites attR (BOP -) and attL (POB') are formed at the time of integration and recombination is therefore achieved using a viral enzyme, ^ integrase, a product of the int gene, and a cellular protein, the IHF host integration factor (heterogeneous dimeric IHF a/B protein encoded by the himA and himD a' *Escherichia coli* genes).

The integrase recognises two specific sequences located within non-homologous DNA: a phage sequence and a sequence from the host chromosome.

Integrase has type I topoisomerase activity: it is capable of cutting Holliday structures and continuing by integration or excision.

Excision requires the presence of the phage protein encoded by the xis gene and also requires the IHF factor.

NB:

The int and xis genes are linked and partially overlap. The PL promoter transcribes xis, but there is another promoter, Pi, which transcribes int independently. The recombination system controlled by int and xis is Rec-independent b)Single specificity site recombination: case of the integration of generalized transduction bacteriophages (bacteriophage u):The bacteriophage iu is an always linear phage which is capable of inserting itself into non-specific sites of the bacterial chromosome.

The 1 genophore contains specific sequences for insertion, but there is no complementary specific sequence on the host. Like transposons, to which it can be assimilated, bacteriophage u carries identical inverted sequences at its ends and contains transposition genes (insertion and excision are recindependent). Like phages, it contains genes encoding phage structural proteins. Because of its particular integration mechanism, the prophage is identical in gene sequence to the free phage (unlike the bacteriophage fy.

When the G segment is reversed on one of the 2 strands, an area of mismatch is created separating two paired regions, or when one of the strands is completely reversed, a paired region (G) is formed between two unpaired regions.

5.5 Transposition phenomena

Site-directed recombination can involve certain sequences of a single DNA molecule by means of internal transposable elements called transposons. These elements are known as "mobile" because transposition leads to a change in their chromosomal location. These phenomena can also lead to the mobility or altered expression of certain genes. There are several types of transposon in bacteria.

The most common are insertion modules (IS elements), which are DNA fragments that can be integrated at more or less specific sites.transposons can move from one site to another, but they never appear in free form.IS elements are not only capable of their own movement, they are also capable of integrating at more or less specific sites.

to move by carrying a sequence located between two of them, this sequence being able to comprise one or more gënes.

These more complex structures are called composite transposons. Composite transposons can be classified into two main categories:

- IS most often inverted (transposon Tn10 with IS10 arm, transposon Tn5 with IS50 arm, etc.), more rarely direct (transposon Tn9 with IS1 arm, etc.);
- Class II transposons (Tnll or TnA family) comprise short inverted sequences (30 to 40 base pairs) flanked at each end by a set of resistance genes and transposition genes (Tnl, Tn3, Tn7, etc.).a transposon is inserted into a receptor site using an enzyme system called integrase. Insertion sites are usually very small (of the order of 5 nucleotides).
- They are duplicated after transposition: the insertion sequence is found at each end of the integrated transposon.

In all cases, a typical structure is formed between the "donor" and "receptor" fragments: the "chi" (x) structure. Single-chain cuts are introduced at the ends of the transposon.

Similarly, a break is created at the insertion site, with the formation of two complementary single strands, from which there are two possible alternatives: conservative transposition and "replicative transposition".

Conservative transposition involves breaking/welding-type steps: it results in a real change in the location of the transposon. Replicative transposition involves making a copy of the transposon and therefore replication: the transposon "mould" remains in place and the copy goes to the receptor site.

Transposition can be simple or lead to more complex rearrangements, such as deletions or inversions.

The loss of a transposon can result in the restoration of a function: this depends on the precision of the transposon excision. Transposition can sometimes favour fusion *i* this property can be used in genetic engineering.

Transposons present on plasmids are sometimes described as conjugative and are often incapable of autonomous replication.

a) Eukaryotic transposons

Transposons also exist in eukaryotic microorganisms: the best known are those found in yeast.

Yeast transposable ëlëments form a hëtërogëne family: the most important are the Ty ëlëments, which are longer than bacterial transposons.

In Saccharomyces cerevisiae, Ty elements can be classified into two main classes based on sequence similarities: Tyl or Ty I (Tyl-15, Tyl09, Ty912) and Ty2 or Ty II (Tyl-17, Ty917). In addition to the Ty (and 5) elements, there are other transposable elements in yeast. The o elements are 341 bp in size and have identical inverted ends (IR) of 8 bp: these elements, of which there are 25 to 30 copies, are associated with tRNA genes.

The rate of transcription of transposons depends on the sexual sign of the strain: it is high in haploi'des a or a and low in diploi'des a/ a. In yeast, Ty transposition occurs at a lower frequency than bacterial transposition (10^{-7} to 10).$^{-8}$

6) Other DNA sequence rearrangements

6.1) Rearrangements in bacteria

Various phenomena of this type have been described:

> Phase variation in Salmonella. Differences in flagellum composition are due to the existence of two types of flagellin, the synthesis of which depends on the H1 and H2 genes. This phenomenon is linked to chromosomal rearrangement. It is not a replicative transposition: there is a simple inversion of a chromosomal fragment between 2 inverted repetitive sequences of 14 bp (IRL and IRR) .

> Antigenic variation in Neisseria. Changes in the antigenic nature of Neisseria gonorrhea,

at the level of the pili, are due to phenomena similar to transposition.

> Antigenic variation in Borrelia. These are variations in the main variable surface protein, which are also due to chromosomal rearrangements.

6.2) Change in the yeast sex sign

The sex-change phĕnomĕnes that occur in some sexed yeast species are due to phĕnomĕnes close to transposition. These are cassette systems. Sex is determined by an a or a gene. Generally, this characteristic is stable, but variations can be observed in certain strains.

Sex depends on the expression of a or a sequences located at a specific site, the sex sign locus (MAT).

6.3) Variations in the trypanosome surface antigen

In the trypanosome (protozoan), phenomena similar to transpositions lead to the appearance of numerous variants for the surface glycoprotein VSG. Each trypanosome can express around a hundred different VSGs: this variability depends on rearrangements in the DNA .

6.4) DNA rearrangements in ciliated protozoa

In Oxytrichia, the DNA of the micronucleus contains internal sequences ranging from 32 bp to several kb (IES: Intemally Eliminated Sequences). These sequences are eliminated during regeneration of the macronucleus and have direct repeat (DR) ends.

7)Cytoplasmic genetic information mixing

7.1- Cytoplasmic heredity

In addition to chromosomal heredity linked to chromosomes or the genophore, there is cytoplasmic heredity due to the presence of extrachromosomal genetic information (plasmids, mitochondria, chloroplasts, "killer" factors, etc.).

In prokaryotic organisms, the genetic information contained in a plasmid can be transferred from one cell to another by transformation or conjugation.

In eukaryotic organisms, transformation by plasmid can also occur. If it is possible, cell fusion leads to the formation of a hybrid cytoplasm and therefore the confrontation of the cytoplasmic genetic information of the two parents. Cytoplasmic heredity is said to be mitotic rather than meiotic.

7.2- Paramecies "killer" factor

During a brief conjugation between a "killer" cell and a sensitive cell, there is no passage of the factor. However, if the conjugation is prolonged, all the cells in the progeny recover the factor.

7.3- Recombination of extrachromosomal DNA

There are phĕnomĕnes of recombination between DNAs derived from organelles. This is known for the DNA of algal chloroplasts (Chlamydomonas reinhardii), or that of yeast mitochondria. This presupposes contact or fusion of the organelles after plasmogamy.

Chapter V

V-Corrected exercises With additional course

1-Course reminder

Analysis of the structural complexity and organisation of the genetic material in prokaryotic, eukaryotic or acaryotic microorganisms is very important for understanding how this molecular machinery works, and the structure-function and structure-activ^ relationships.

Nucteic acids are the ëlëments of the gënëtic material, they are compo- sës of liiK\iary chains of nucteotides Нёз between them by phosphodiester bridges.

Nuckotides are formedës by a purine or pyrimidine base Нёс to a pentose (Ribose or dësoxy-ribose) which is itself Нё to a phosphate group .

Dësoxyribonucleic acid (DNA) is formed from 2-ck'sowribose and ribonucleic acid (RNA) from ribose.

Nuckosides are formed by the association of a nitrogenous base with a pentose via an N-glycosidic bond (Figure 2) between carbon 1 of the ribose and nitrogen 1 of the pyrimidine base or nitrogen 9 of the purine base.

The assembly of the chains forms a double helix in which each strand is complementary to the other and antiparallel: one chain has a polar^ 5'-> 3' and the other 3'-> 5'.

For a doniK'e species, the overall base composition of the DNA is a constant: this composition can be expressed by the ratio A+T/G+C. This ratio varies from one species to another, but it is constant and characteristic for a given species.

The kilobase (kb) is usually used as a иилё measure of the number of nucteo- tides. The kilobase corresponds, for bicatëary nucic acids, to 10^3 base pairs (bp). or, what amounts to the same thing, nucteotides. We consider that 1 kb corresponds to a molecular mass of 0.7.106 Da (1pb gives approximately 660 kDa) and a size of 0.35 |im.

We find approximately 10 base pairs per turn of the turn; The spacing between two consëcutive bases is 0.34 nm (3.4A°)The pitch of 1^ëHcc is 3.4 nm (34A°)and the dia- metre of PЬёHcc is 2 nm (20A°).

A simplified structure of double-stranded DNA, with its two complementary and anti-parallel choanes, is shown in Figure 3.

In prokaryotic organisms, the main genetic material is a single double-stranded DNA structure called a chromosome or genophore, located in an area of the cell called the nucleus region or nucleoi'de. The chromosome is circulatory, meaning that it is closed. The DNA is complexed with Mg^{2+} or Ca^{2+} cations and with proteins (protamines) and others (HLP: Like histon protein).

The main genetic material of eukaryotic organisms is made up of chromosomes whose basic element is, as in prokaryotic organisms, a DNA double helix. The chromosomes are enclosed

in a defined cellular structure called the nucleus. The nucleus is surrounded by a nuclear membrane which contains pores and is linked to the cell's internal membrane system.
There are two categories of DNA-bound proteins: histones and non-histone proteins (HMPs). The basic structural element of the chromosome is the nucleosome.
There is genetic material outside the genophore or chromosomes. This material can represent 0.1 to 1 and sometimes more than *10%* of total cellular DNA. "Extra-chromosomal genetic material" includes DNA from organelles (mitochondria and chloroplasts), plasmids (bacteria and yeast), "killer" factors (yeast and paramecium), as well as the genetic material of viruses (bacteriophages).

2-Requisites

To ensure a good understanding, the student is asked to have an idea of :
> Diversity of the microbial world.
> The structure of nucleotides.
> The organisation of genetic material.

3-Objectives :

> Learn the basic concepts concerning the structure and organisation of microbial gënëtic material.
> Differentiating between prokaryotic and eukaryotic microorganisms in terms of structural and molecular complexity.
> Familiarise yourself with scientific terms and calculation formulas.

Series n :1 "Structure and organisation of genetic material".

Exercise 1:

1-DNA paration;2-DNA hëlicase;3-Homologous recombination;4-DNA primase;5-DNA polymërase;6-S-phase;7-Origin histone chaperone (ORC);8-Protëine S SB (Single-Strand DNA-binding);9-Base excision-repair;10-Reproduction fork;11-Nucleotide excision-repair;12-RNA primer;13-DNA topoisomerase;14-Slip factor;15-Replication root;16-DNA ligase;17-Rëunion of non-homologous ends;18-Tël omërase;19-Precose strand;20-Mutation; 21-Late strand;22-Recognition complex.

Give each of the following definitions a term from the list above.

1-Phase of the eukaryotic cell cycle during which DNA is synthëtisë.
2- Enzyme that lengthens the tël omëres, the rëtiative nucleotide sequences found at the end of eukaryotic chromosomes.
3-One of the two newly Бэгтë strands of Иоиуë DNA has a replication fork. It is synthëtisë de fagon continuous in the 5'-3' direction.
4-Short fragment of RNA synthëtisë on the late strand during DNA replication and then 'limin' .
5-Genetic exchange between a pair of identical or nearly identical DNA sequences, usually those located on two copies of the same chromosome: mismatched or homologous chromatids.
6- A means of repairing broken bicatëary DNA, which links the 2 ends with little respect for the homology of the sëquences.
7-Protëine complex that encircles the DNA double hël ice and binds to the DNA pol ymërase. keeping it firmly Hë to the DNA as it dëpl ace.
8-1 rel'ment containing information that controls a discrete hëditary character.
9 Enzyme that binds to DNA and reversibly breaks a phospho-diester bond in either strand.
10- Collective term given to the enzymatic processes that correct differential changes

affecting the continuity or sequence of a DNA molecule.

11-Large multimeric protein structure linked to the DNA that replicates eukaryotic chromosomes throughout the cell cycle.

12-One of the two newly formed strands of DNA found has a replication fork. It is formed of discontinuous segments that are later covalently linked.

13-Y-shaped region of a replicating DNA molecule where the two brother strands are formedës.

14-Enzyme which opens the DNA hedge by sëparating the single strands.

15-Particular DNA sequence of a bacterial or viral chromosome at which DNA replication begins.

16-Enzyme that joins two adjacent strands of DNA

Exercise 2:

If the GC content of a DNA molecule is approximately 56%, what are the percentages of the other four bases (A, T, G and C) in this molecule?

Exercise 3:

By extracting the DNA from coliphage ф X174, we found that its composition is 25% A, 33% T, 24% G and 18% C.

> Does this make sense in terms of Chargaff's rules?

> How do you interpret this result?

> How could such a phage represent its DNA?

Exercise 4:

A DNA segment has 24 nitrogenous bases such as

> Determine the number of each nitrogen base.

A segment 34 A° long and 12 hydrogen bonds > Determine the number of each nitrogen base.

A DNA segment has 38 phosphodiester bonds such that and the A+T ratio is 2 times the Chargaff coefficient.

> Determine the number of each nitrogen base.

Exercise 5:

The Escherichia coli chromosome has a molecular mass of 2.5 x 109 Da. ;

Calculate :

> The number of base pairs

> Length

> The number of turns of the helix of this molecule

Exercise 6:

The тоlёсиlе of *Escherichia coli* DNA comprises approximately 2.6.106 base pairs and weighs $2.66.10^{-15}$ g. Knowing that the pitch of the double hëlice of the DNA is 3.4 nm and com- prises 10 base pairs ;

> Calculate the mass of a segment of DNA 0.1 nm long;

- in grams.
- in daltons.

Exercise 7:

A cell synthesises 5000 different polypeptides, each of which has an average weight of around 24000. Knowing that the average molecular weight of an amino acid is 120, that of a nucleotide 300 and the length of a nucleotide 3.4 A

> What is the total length of the DNA encoding these polypeptides?

> Calculate the molecular weight of this double helix DNA segment

Exercise 8:

109 Escherichia coli cells are cultured (generation time: 30 minutes).

> After how long will there be around 1 billion cells?

Knowing that the DNA molecule *of Escherichia coli* contains approximately 2.6.106 base pairs;

> Calculate the number of base pairs that should undergo replication per second, assuming that there is only one replication point (we will assume that the replication time of the DNA molecule is equal to the generation time).

> How far should the DNA polymerase travel per minute, assuming that there is only one replication point?

> How long would the DNA molecules from 109 *Escherichia coli* cells be if they were placed end to end?

Solution

Exercise 1:

1-6

2-18

3-19

4-12 5-3 6-17 7-20 8-14 9-13 10-1 11-7 12-21 13-10 14-2 15-15 16-16 **Exercise 2 :**

Chargaff's rules stipulate that the DNA of any cell must have a 1:1 ratio (base pair rule) of pyrimidine and purine bases and, in other words, in a given sëquence there will be as much guanine as cytosine and as much adënine as thymine.

Since C always pairs with G in a DNA duplex, their molar percentages must be equal. The molar percentage of G is therefore, like C, 28%. 100. The molar percentages of A and T correspond to the remaining 44%. 100%. Since A and T always match, their mole percentages are half this value: 22%. 100.

So if we recapitulate the proportions of each base

C=G=28

A=T=22

Exercise 5:

1-Total number of base pairs

Taking into account the molecular weight of a base pair equivalent to 660 Da,

The number of total base pairs is calculated as follows:

N(bp)= Molecular mass/Mass of an AN base pair :

N(bp)=2.5 109/660=**3.78 10 bp.**^{6}

2-Length

The distance or length of a nucleotide is equivalent to 0.34nm and/or 3.4 A0, so the length of all base pairs is given as follows:

L=Number of base pairs x Length of an AN base pair :

L=3.78 10^6 x 0.34=1.2 10^6 nm=1**.2cm.**

Exercise 7:

1-Calculation of total amino acids :

N(AA)=(Average weight of polypeptide /Weight of one amino acid) x number of polypeptides

(N(AA)=(24000/120)**x5000=10^6 AA**

2-Length :

Taking into account that amino-acid is encoded by three nucleotides each having an equivalent length of 0.34nm:

L(AA)=(Length of a nucleotide x 3)x Number of amino acids

L(AA)=$0.34 x 3 x 10^6$ **=1.02cm** 10^6 nm=1.02cm

3-The molecular weight of this DNA segment

The molecular weight of a nucteotide is ëquivalent to 300 Da,and each three nucleotides constitutes a single aттë acid, and the total calcиlë acid number is about 106 AA.Such a number is tlieorically equivalent to the number of constituent nucleotides. Similarly, it must be taken into account that the DNA structure is duplex with the presence of two complementary and antiparallel chains.

The weight is calculated as follows

Weight=(Weight of one nucteotidex3)x(Number of amino acids) x2(Double strands)
AN :
P=$3x300x10^6$ x2=1.8 $x10^9$ Da or **1.8 10^6 kDa.**

1-Course review

Replication (or duplication) is the mechanism that enables the reproduction of DNA molecules, which, except in special cases (single-stranded DNA viruses), are organised in a double helix. This structure is opened up, forming a "replication fork" and polymerising new strings of nucleotides that are complementary to the existing strings. Replication is semi-conservative (Experience of Messelson and Stahl, 1957) and polymerisation, which is carried out by enzymes called DNA polymerases, cannot take place from a simple mixture of nucleotides.

Replication involves a large number of proteins, including a multi-protein complex called the replisome, which has three distinct phases (initiation, elongation and termination).

a) Initiation

Replication begins with the opening of the double helix at a specific site (the initiation point *ori C in* prokaryotes or ARS (autonomous replicative sequences) in yeasts (eukaryotes).The separated strands are stabilised by proteins called "single strand binding protein" (SSBP or SSB=HDP: helix destabilising protein=DBP: DNA binding protein): these are also tetramëres.

b)Elongation :

DNA polymerase III, or DNA pol 6 in eukaryotes, does most of the elongation work. The RNA initiator strand is eliminated and the missing part is completed via the polymërasic activity of DNA polymerase.

Replication therefore occurs asymmetrically on the two chains, forming short discontinuous fragments in the late strand. "Okazaki fragments".

c) Termination :

In *E. coli*, a fork meeting will take place at a point diametrically opposed to the replication initiation site. In eukaryotes, once the DNA has been duplicated, it is reorganised into nucleosomes to return to its initial conformation (chromatin). DNA replication in eukaryotic organisms is accompanied by a process of histone duplication. Replication is bidirectional and generally takes place from several origins.

For the replication of extrachromosomal DNA, mitochondrial DNA and plasmid DNA are replicated using a mechanism similar to that of chromosomal DNA. These DNAs often contain a single repticon.

The replication of gëtic material and the intracellular multiplication of vi- rus (Bacteriophages) largely use the host's enzymatic ëquipments. The mechanism differs according to the nature of the 1 DNA or RNA.

Prerequisites :

It is strongly recommended that the student has an idëe at

> The structural difference between prokaryotes and eukaryotes.

> The basis of the replication process

> The enzymes involved and the usefulness of duplicating genetic material.

Objectives

> To understand the vital needs of microorganisms, giving details from a molecular and genetic point of view.

> Learn about the different levels of DNA compaction and/or supercoiling in eukaryotes.

> Developing intellectual skills in terms of reflection and analysis.

Series n :2 "Replication of genetic material".

Exercicel :

You syntlictisev DNA onto the 5'-ACCT-TACCGTAATCC-3 'template from an upstream primer. You add three dësoxynuclëotides, and leave the fourth desoxynucleotide "dësoxycytosine triphosphate", from the reaction mixture.

> Which DNA will you have? Draw a picture.

You synthesise DNA from the same template and with the same upstream strand, but instead of just using dësoxvthvmidine triphosphate, you add some to the other three dësoxvnuclëosides triphosphates,a mëlange ëquivalent to the replication inhibitor, didesoxythymidine triphosphate,and dësoxvthvmidine triphosphate,

> What DNA would you make? Draw a picture.

Exercise 2:

A nuclCosome is 11nm long and contains 146bp of DNA (0.34nm/pb).

> What compaction ratio (DNA length to nucleosome length) was obtained by wrapping the DNA around the histone octamëre?

> Taking into account the additional 54bp segment of CtDNA linking the nucleosomes, what is the degree of compaction of the "string of pearls" DNA compared with totally CtDNA?

Exercise 3:

Knowing that the *E.coli* gCnome has 4.6 106 base pairs, that the oxidation of one molecule of glucose yields around 30 ëenergy-rich phosphate bonds, and that glucose has a molecular mass of 180 daltons, and there are 6 10^{-23} daltons per gram ;

> Approximately how many ëenergy-rich bonds are used to fold the *E.coli* chromosome?

> How many molecules of glucose will *E.coli* need to consume to have enough energy to copy its DNA once?

> What does this mass of glucose represent in relation to the mass of *E coli,* which is around 10 g?$^{-22}$

Exercise 4:

By analysing two yeast colonies, we have found that they each contain 100,000 cells descended from a single yeast cell, originally located somewhere in the middle of the cluster. When the *ADE2* gene is located at its normal chromosomal locus, the protein it encodes is expressed and the colonies are white. However, when it is located near a telomere, the *ADE2* gene is inactive, resulting in red colonies with white sectors. However, the ADE2 gene is located near telomeres in both red and white sectors.

> Explain why white areas have formed around the edge of the colony.
> Based on the existence of these white sectors, what can you conclude about the propagation of the transitional state of the ADE2 gene from my own cell to the daughter cells?

Exercise 5:

Like all organisms, the T4 bacteriophage encodes an SSB protein, essential for suppressing secondary structures in single-stranded DNA upstream of the replication fork. The T4 SSB protein is a stretched monomeric protein with a molecular weight of 35,000 Da. It binds tightly to single-stranded DNA but not to double-stranded DNA. Binding is saturated when the DNA to protein weight ratio is 1:12. In the presence of excess single-stranded DNA (10 jag), virtually no binding is detectable with 0.5 Lig of SSB protein, whereas almost complete binding is observed with 0.7^g of SSB protein.
> At saturation, what is the ratio of single-stranded DNA to SSB molecules (the average molecular mass of a nucleotide is 330 daltons)?
> When SSB protein binding to DNA reaches saturation, is it likely that adjacent SSB monomers will be in contact? (Suppose that an SSB monomer extends 12 nm along the DNA when bound to it, and that the base spacing in single-stranded DNA after binding is the same as that in double-stranded DNA (i.e. 10.4 nucleotides per 3.4 nm).
> In your opinion, why does the binding of the SSB protein to single-stranded DNA depend so strongly on the quantity of SSB proteins?

Solutions to some exercises

Exercise 2:

The nucleosome represents the first level of DNA compaction in eukaryotes. It is an octamer of histones on which a DNA segment of approximately 146 base pairs is wound in two turns.

1-The compaction ratio :

It represents the degree of compaction děcriving the passage from the double-stranded DNA fome to the nucleosome form:

R=(146 bp x 0.34 nm/bp)/(11 nm) = **4.5.**

2-The degree of compaction of "string of pearls" DNA compared with fully expanded DNA:

The second level of compaction involves the presence of 6 nucleosomes linked together by the DNA linker, forming a string of pearls known as a solenoi'de.

Counting the additional 54 bp of interucieosomal DNA (146+54=200 bp), the compaction ratio of the DNA in the pearl necklace is calculated as follows:

R=(200 bp x 0.34 nm/bp)/(11 nm + {54 bp x 0.34 nm/bp}) = **2.31.**

The first level of compaction represents only 0.023 p. 100 (2.31/10000) of the total condensation that occurs during mitosis.

Exercise 3:

1-Number of energy-rich bonds :

As both strands of the *E. coli* genome must be copied, a total of 9.2 x 106 nucleotides must be polymerised. The polymerisation of triphosphate nucleosides in DNA consumes two high-energy phosphoanhydrogen bonds for each nucleotide added:

The nucleoside triphosphate is hydrolysed to add nucleoside monophosphate to the growing strand, and the liberated pyrophosphate is hydrolysed to phosphate:

The number of high-energy phosphate bonds hydrolysed during each replication cycle will be :

N=9.2 x 106 x2=1**.8x107 bonds**

2-Number of glucose molecules to be consumed :

As each glucose can provide 30 high-energy phosphate bonds, we will have as a number of glucose molecules needed to provide enough energy for one replication cycle as follows:

$$N(molecules)= (1.8^{x} 107/30)= \mathbf{6 \text{ x } 105 \text{ molecules}}$$

3-Mass calculation of glucose molecules :

$M=(6^{x}\ 105 \text{ molecules})^{x}\ (180 \text{ Da/molecule})^{x}\ (1 \text{ g}/6^{x}\ 10^{-23} \text{ Da})]=1.\mathbf{8x10\ g^{-16}}$

This quantity of glucose is approximately 0.02 p. 100 of the mass of an *E. coli* cell.

$P(\%)=(1{,}8 \text{ x } 10^{16} \text{ g glucose}/1 \text{ x } 10^{12} \text{ g } E.\ coli)=\mathbf{0{,}02\%}$

Exercise 5:

1-The 1:12 ratio of the weight of SSB protein nucleotides corresponds to a ratio of **8:8:1** nucleotides of SSB molecules :

$$\frac{\text{Nucléotides}}{\text{molécules SSB}} = \frac{\text{1Da Nucléotide}}{\text{12 Da SSB}} x \frac{\text{35000 Da SSB}}{\text{molécules SSB}} x \frac{\text{1 Nucléotide}}{\text{330 Da nucléotides}}$$

2-Since there are 10.4 nucleotides per 3.4 nm of single-stranded DNA, the 8.8 nucleotides would extend about 3 nm. (If the single-stranded DNA had been fully extended, it would extend about twice as far. The SSB molecule length of 12 nm suggests that at saturation the SSB proteins are in contact with each other and probably chew each other up considerably.

3-The absence of significant binding at low concentrations of SSB with the presence of quantitative binding at 14-fold higher concentrations suggests that the SSB protein binds DNA cooperatively. In essence, cooperation means that the binding of one monomer is made easier for additional monomers to bind. If monomers overlap when bound, as suggested by the calculation in part 2, cooperativity probably occurs because each monomer has two binding sites; one for DNA and one for other monomers. The binding of the first monomer to DNA will be weak because it can only bind through its DNA binding site.

By binding to adjacent linked monomers, the following monomers can make use of both of their binding sites.

1-Course reminder

A mutation is a hërëditary modification of an individual's gënëtique material,in the absence of confrontation with a ëtranger's gënëtique material.This modification may or may not be reflected at the phënotypic level(visiblеs).

We distinguish five characteristics: Spontaieity, Discontinuity, Raretë, Indëpendence and specificity and Stability.

Its frequency of appearance varies between 10^{-4} and 10^{-12} , but more generally, between 10^{-6} and $10-^{9}$ per generation. It is a spontaneous phenomenon.

In principle, a mutation occurs at random. The change brought about by the mutation is hereditary, i.e. it is passed on to descendants. The mutation can be reversed by a new mutation. Many spontaneous mutations are letal.

The mutation rate is the probability of a mutant appearing in a population during a given time interval. This time interval is generally the generation time for unicellular organisms.

The rate of mutation depends on the nature of the mutation (species, gene affected), the mutagen and its dosage (in the case of induced mutations) and possibly the surrounding physico-chemical conditions.

The mutant frequency depends on the mutation rate, the growth parameters of the wild and mutant strains and time. Its value is deduced from the slope of the curve frequence=f(n ;nbre of generations)(straight line).

Salvador Luria and Max Delbruck developed a test called the fluctuation test, which implies the random nature of natural mutations, independent of any inducible consequences of natural selection.

There are many different types of mutation, depending on the location, size of fragment affected and consequence.

Mutations can be classified into nuclear (or chromosomal) mutations and cytoplasmic mutations depending on whether they affect chromosomal genetic material (chromosomes or genophores) or extrachromosomal material (plasmids, mitochondria, etc.).

Depending on the size of the genetic material involved, a distinction is made between:Genomic mutations,Chromosomal mutations. Genomic mutations. These are found in both prokaryotic and eukaryotic organisms. They are mutations affecting a single gene (intragenic mutation; point mutations and non-point mutations).

Depending on the number of base pairs touchedë we find point mutations and mutations by reshuffling.

Point mutations involve a single base pair. There are the following cases: Substitution (transition and Transversion), Dëlëtion, Insertion and Modification.

Shuffle mutations involve more or less extended chromosome (or chromatid) ëments ranging from a few base pairs (microshuffles) to polygënic fragments ;Deletion (terminal endochromosomal, exochromosomal intragënic microdëlëtions),Inversion, Translocation (homozygous and heterozygous),and duplication.

2-Prerequisites :

L^tudiant est appe1ë a connaitre la notion de :

> Mutation and mutant.
> The propriëtës and cara^ristics of a mutation.
> Types of mutations.

3-Objectives :

> Learn to reduce the mutation rate, frequency and probability of occurrence.
> Understanding the notion of auxotrophy and the acquisition of new gënëtic ciracteres.
> Draw up a compliant biochemical pathway based on an in-depth analysis of the data.

Series n :3 "Mutations

Exercise 1:

The *dcm* m6thylase adds a тёЛу1е group to the second cytosine in the CCAGG sëquence in *E. coli* K-12. The cytosines in this sëquence in the *lacI* gene are susceptible points for transition mutations .

> Sugërez two mëthodes gënëtiques for dëterminer if the mutation frequency is ëlevëe at this position is due to 5-mëthyl cytosine.

Exercise 2:

Wild-type *Salmonella typhimurium* cells are killed by phage F0. A Luria-Delbruck fluctuation test has ë1ë геяНзё to dë determine the mutation rate of resistance to phage F0;Twenty tubes of medium have ë1ë inoculated with *S. typhimurium* cells and cultures have ë1ë grown to 108 cells / ml. A 0.1 ml ë sample of each culture was then ë1ëtalë onto plates with the FO phage (Table 3). The results are shown below.

> Using the Poisson distribution, determine the mutation rate in FOR.

Table 3: Results of the effect of FOR phage on the culture of *S.typhimurium*

Culture	1	2	3	4	5	6	7	8	9	10	11	12	13	14	15	16	17	18	19	20

Mutant F0 .[R]	59	0	1	17	0	6	0	0	0	3	1	0	0	0	1	128	0	2	0	11

Exercise 3:

Proline transport mutants appear spontaneously in *Salmonella* culture at a frequency of 10^{-6} per cell. Azetidine-resistant mutants appear spontaneously at a frequency of around 10^{-5} per cell.

> If these are two independent events, what is the probability that a single cell will acquire both mutations?

> If 100 proline transport mutants are tested and all are equally resistant to azetidine, how can you explain the observed frequency of each type of mutant?

Exercise 4:

DHP is transported into the cell by proline permëase and is incorporated into proteins, resulting in defective proteins and consequent cell death. Spectinomycin binds to ribosomes, inhibiting protein synthesis and leading to cell death. After treatment with the intercalating agent ICR-191, the frequency of DHP mutants^{r} is around 10 to 4 per cell but the frequency of Sprr mutants is around 10 to 9 per cell.

> Suggest an explanation for the difference in mutation rates between DHP^{r} and Sprr and propose a genetic experiment to test your idea.

Exercise 5:

Wild-type *E. coli* bacteria are unable to grow in the presence of decanoic (fatty) acid (Dec-). A Luria-Delbruck fluctuation test was performed by plating 108 cells obtained from 10 independent cultures onto plates of decanoic acid and, simultaneously, plating 108 cells from a single culture onto 10 plates of decanoic acid (Table 4). The results obtained are presented below:

Table 4: Results of culturing wild-type *E. coli* in the presence of 1'-decanoic acid

Indian culture-hanging	Number of colonies	Culture (Only one)	Number of colonies
1	22	1	17
2	18	2	24
3	19	3	23
4	24	4	26
5	20	5	20
6	23	6	22
7	21	7	21
8	22	8	24
9	21	9	20
10	17	10	19

Based on these results, is the Dec + mutation random or adaptive?

Exercise 6:

You have infected an *E coli* culture with a virulent bacteriophage. Most of the cells are lysed, but some survive: 1 x 10^{-4} cells in your sample.

> Are the resistant bacteria caused by the bacteriophage infection, or did they already exist in the bacterial culture?

Earlier, in another experiment, you spread a diluted suspension *of E coli* on a solid medium in a large petri dish and, after observing that around 103 colonies were growing, you made a replica of this dish on three other dishes. You realise that you can use these dishes to test your two hypotheses.

You place a suspension of bacteriophages on each of the three replica dishes so that the bacteria can be infected.

> What should you look out for if the bacteriophage creates resistance? What should you look out for if bacterial resistance already exists?

Exercise 7:

Three different proline auxotrophs were streaked onto minimal proline-free plates and incubated at 22°C, 30°C or 42°C for two days (Table 5). The results are presented in the table below:

Table 5: Incubation results for the three auxotrophs at different temperatures.

Double mutant		**Growth on minimal medium a:**
		22°C30°C42°C
pro-1	pro-2---	
pro-1	pro-3-+-	
pro-2	pro-3---	

> For each strain, indicate whether the pro-mutation is a null mutation, a temperature-sensitive mutation or a cold-sensitive mutation.

Exercise 8:

Mutations are introduced into the *E. coli* genome at the rate of one mutation every 109 base pairs per generation. Imagine that you start with a population of 109 *E. coli*, none of which carries a mutation in your gene of interest, which is 1000 nucleotides long and essential for neither bacterial growth nor survival.

At the next generation, once the population has doubled in number ;

> What fraction of cells on average will carry a mutation in your gene?

> Once the population has doubled again, what do you think the frequency of mutants in the population will be?

> What will be the frequency after a third doubling?

Exercise 9:

The DNA and corresponding amino acid sequence of part of the wild-type *trpA* gene is shown below. The DNA sequence of three different *trpA* mutations is shown directly below the corresponding region in the wild-type sequence (the DNA sequence of the coding strand is shown - i.e. TTG = UUG = Leu).

```
209 210 211 212 213 214 215 216 217 218 219 220 221
TTG CAG GGA TTT GGT ATT TCC GCC CCG GAT CAG GTA AAA
Leu Gln Gly Phe Gly Ile Ser Ala Pro Asp Gln Val Lys
 *
TAG CAG GGA TTT GGT ATT TCC GCC CCG GAT CAG GTA AAA
                     *
TTG CAG GGA TTT GGT GTT TCC GCC CCG GAT CAG GTA AAA
TTG CAG GGA TTT GGT ATT TCC GCC CCG ATC AGG TAA AA-
```

> Using the codon table, show the amino acid sequence of the protein produced in each of the three *trpA* mutants.

Exercise 10:

Four single mutant strains of *Neurospora* cannot grow on minimal medium unless supplemented with one or more of substances A-F. In the table below growth is indicated by + and no growth by 0. In addition, both species 2 and 4 grow if E,F, or C and F, are added to the minimal medium (Table 6).

> Draw a biochemical pathway consistent with the data, involving each of the six metabolites, where mutant blockage occurs in each of the four strains.

Table 6: Results of *Neurospora* growth in the presence of several substances

Strain	A	B	C	D	E	F
1	+	0	0	0	0	0
2	+	0	0	+	0	0
3	+	0	+	0	0	0
4	=	0	0	0	0	0

Solutions to some exercises

Exercicel :

The precise determination of spontaneous mutation rates and frequencies can contribute to our understanding of these processes and the enzymatic pathways that handle them. The тё-thodes used to calculate mutations are based on the expansion of mutant clones by Luria and Delbruck and extended by Lea and Coulson. Mutation frequency and mutation rates are highly correlated with each other.

Several tests are used to measure the frequency and mutation rate in a particular gene pool. Some of the tests are as follows:

-Avida Digital Evolution platform and

- Fluctuation test.

On the other hand, to answer the question of whether the mutation frequency is high at the level of the second cytosine of the CCAGG sequence in *E. coli* K-12, we can simply do the following;

1-Dosing the mutation frequency in a dcm mutant.

2-Determine the frequency of *lacI* gene mutations in a mutant with a modified *dcm* recognition sequence (e.g. mutated to CAAGG).

Exercise 3:

1-The probability of two independent events is equivalent to the product of the frequency of their occurrence:

$$P=10^{-6} \times 10^{-5}=10^{-11}$$

2- If all proline transport mutants are also resistant to azetidine, proline transport mutants represent $10^{-5} / 10^{-6}$ = **10%** of azetidine-resistant mutants. Thus, there must also be other types of mutation that can also lead to resistance to Azetidine.

Exercise 4:

- ICR is an intercalating agent and therefore causes frameshift mutations which generally result in a so-called null phenotype (no phenotypic expression).

The relatively high frequency of DHP mutants^r suggests that this gene is not essential, as it is easily disrupted by shift mutations.

However, the low frequency of Spc mutations^r suggests that ICR does not stimulate mutations in this gene, probably because the gene is essential.

The few mutants obtained are probably due to spontaneous mutations that do not completely per- turbate the gene. To confirm this idea, it would be advisable to compare the frequency of spontaneous mutations in Spr^r with the frequency of mutations in Spr^r generated by ICR.

Exercise 5:

The variance between the number of Dec + colonies in individual cultures is equal to that observed with several samples from the same culture. These results suggest that the phenotype is probably not due to a random, spontaneous mutation. (A random mutation that would show much greater variance between individual cultures). So, the simplest interpretation of these results is that the Dec + mutation is an example of "adaptive mutagenesis".

Exercise 6:

- For either hypothesis, we need to observe around 10 surviving colonies per dish. If the bacteriophages induce resistance, the surviving colonies should appear at random positions on each replicate plate.

-On the other hand, if the mutations preexist, the resistant colonies should appear at the same positions on each of the three replicates. In experiments of this kind, the surviving colonies appear at the same locations, indicating that the mutations pre-exist in the population.

Exercise 7:

By analysing the results obtained we can deduce that :

- The null mutant will be inactive at all temperatures.
- The temperature-sensitive mutation will be inactive at 42°C but active at lower temperatures.
- The cold-sensitive mutation will be inactive at 22°C but active at higher temperatures.

So, when temperature and cold sensitive mutations are combined, the phenotype strain will grow at 30°C but not at 22°C or 42°C so

pro-1 = temperature sensitive (heat)

pro-2 = nil (unconditional)

pro-3 = sensitive to cold.

Exercise 8:

1-In the initial population, there are 106 copies of your 1000 bp gene, i.e. a total of 109 bp which must be copiedë when the population doubles.

With a mutation rate of one mutation every 109 bp per generation, we can expect one copy of our 1000 bp gene to carry a mutation (one mutant cell) in the population of 2 x 106 cells, representing a frequency of 5 x 10^{-7} [1 mutant cell/(2 x 106 cells total)].

2-After the first doubling, there are 2 x 10^{6} copies of your 1000 bp gene, i.e. a total of 2 x 109 bp that need to be copiedëes during the next population doubling.

With the same mutation rate, we now expect 2 mutant copies of the gene to be generated, corresponding to a frequency of new mutants of 5 x 10^{-7} [2 mutant cells/(4 x 106 total cells)]. However, the frequency of mutations in the population is higher because the mutant cell generated in the first doubling divides to produce two mutant cells in the second generation, giving an overall mutant frequency of 10- 6 [4/(4 x 106)].

2-After the second doubling, there are 4 x 10^{6} copies of your 1000 bp gene, making a total of 4 x 10^{9} bp to copy. We therefore expect 4 mutant copies of the gene to be generated, giving a frequency of new mutants of 5 x 10^{-7} [4/(8 x 106)] in the third generation.

The four mutant cells present after the second doubling will also double to produce 8 mutant cells; the overall frequency of mutants is therefore 1.5 x 10- 6 [12/(8 x 10^{6})].

This exercise illustrates a key difference between mutation rates and mutation frequencies. Mutation rates are constant, whereas frequencies increase as the cell population grows.

1-Course reminder

Mutagenesis is a process by which the genetic information of an organism is modified, resulting in mutation. This can occur spontaneously in nature or following exposure to so-called mutagenic agents.

Directed mutagenesis is a high-quality method involving the study of genes, the genome in a context of functional analysis, understanding the molecular mechanism of biodiversity at different organisational levels.
There are many different mutagenic agents, and their use depends on the objective of the experiment being set up. We distinguish between the following agents;

- Physical agents (ultraviolet radiation, ionising radiation, etc.).
- Chemical agents (base analogues, base transformation agents, alkylating agents, nitroso compounds, metals, mycotoxins, hydrocarbons).
- Other agents that disrupt the functioning of genetic material (Intercalating agents, Inhibitors of nucleic acid synthesis, Inducers of polyploi'die).

Repair systems have a downstream action and are active following the failure of control points and post-trascriptional and post-translational regulation processes in order to minimise the extent of the consequences and to avoid the definitive and irreversible loss of control.
To date, four major repair systems have been described;

- Repair by direct reversion of the lesion:

J Photoremediation after ultraviolet (UV) mutation involving the presence of a particular photoreactivation enzyme, "photolyase" (or PR enzyme) which is activated by blue light (300 to 600 nm).
J Desalkylation: alkylguanine-DNA alkyltransferase (product of the ada gene) which is capable of desalkylating alkylguanine without breaking the phosphodesoxyribose backbone.

- Homology-dependent repair systems:

J Excision repair
❖ Base excision repair (BER)
❖ Nucleotide excision repair (NER) system
J MMR mismatch repair (post-replicative)
- Rëparation of double strand breaks (DSB)
❖ Rëparation via the NHEJ system
❖ Repair by homologous recombination
- SOS (Save Our Selves) repair or by pass.

2-Pre-requisites :

J Knowing how to distinguish between a spontaneous mutation and an induced mutation
-*S* Understand the concept of mutagenesis and mutant agents.
-*S* Have a general idea of the control points.

3-Targets

-*S* Know the advantages of random and directed mutagenesis.
J Understanding the consequences of mutagenic agents, with concrete examples.
J Study protection strategies and the effectiveness of repair systems.

Series n :4 "Mutagenesis and Repair Systems".

Exercise 1:

Mutagenesis with UV light causes a variëtë of mutation types, but the ICR181 mutagen specifically causes frame-shift mutations. A large number of *Klebsiella aerogenes putP* mutations were obtained with both mutagens. Approximately 5% of *putP* mutants obtained following UV mutagenesis had detectable CRM, whereas none of the mutants obtained following ICR-181 mutagenesis had detectable CRM.
> What does the presence of CRM mean?
> Sugërez a probable explanation for these results.

Exercise 2:

You and your manager are trying to isolate mutants in *E.coli* using UV light as a mutagen.

To obtain a very large number of mutants, you realise that you need to use a dose of irradiation that kills 99.99% of bacteria.

You worked through the night and obtained much more consistent results than those obtained in the morning by your consultant, who, moreover, needed irradiation doses 10 to 100 times higher to obtain the same Mtalit^.

> How can we explain this difference in results?

Exercise3 :

Mutagens such as N-methyl-N'-nitro-nitrosoguanidine(MNNG) and methylnitrosource (MNU) are powerful mëthylating agents and are extremely toxic to cells.

Bacteria were plated first in a low concentration of MNNG (1 |ig/ml) for 90 min and then in a new medium without MNNG. At different times during and after exposure to the low dose of MNNG, samples of the culture were treated with a high concentration (100 |ig/ml) of MNNG for 5 min and tested for viability and mutant frequency.

As shown in Figure 38, exposure to a low dose of MNNG increases the number of survivors and temporarily decreases the frequency of mutants among surviving cells.

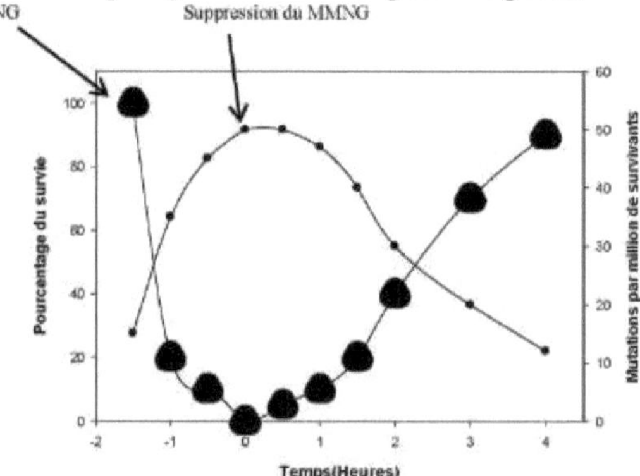

Figure 38: Adaptive response *of E.coli* to a low dose of MMNG

As Figure 39 also shows, this adaptive response to low-dose MNNG is prevented if chloramphenicol, a protein synthesis inhibitor, is added to the incubation medium.

> Does the adaptive response *of E.coli* to a low level of MNNG require the activation of a pre-existing protein or the synthesis of a new protein?

> In your opinion, why is the adaptive response to a low dose of MNNG short-lived?

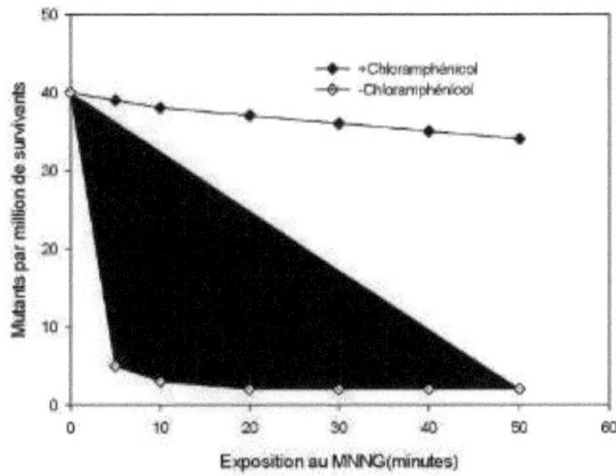

Figure 39: Effect of chloramphenicol on the adaptive response *of E coli*

Exercise 4:

As well as killing bacteria, UV light causes mutations. You have measured the frequency of UV-induced mutations in wild-type *E. coli* and in strains defective in either *the uvrA* or *recA* gene (Table 7). The results are presented in the table below.

Table 7: Frequency of UV-INDUCED mutations in different *E. coli* strains

Strain	Survival(per 100)	Mutations/10^{10} survivors
Wild type	100	400
Rec A	10	1
UvrA	10	40000

Surprisingly, the frequencies of UV-induced mutations differed specta- cularly in these strains.

Assuming that the products of the *recA* and *uvrA* genes participate in different repair pathways for UV damage.

> Indicate which route is most prone to errors.

> Which pathway predominates in wild-type cells?

The error-prone pathway is the result of erroneous incorporation of nucleotides opposite unrepaired altered sites by specialised DNA polymerases.

One of them tends to incorporate an adenine opposite a pyrimidine dimer.

> Is this a good strategy in the face of UV damage?

> Calculate the frequency of base changes (mutations) when an A is inserted versus random incorporation (each nucleotide with the same probability) for *E. coli*, in which pyrimidine dimers are approximately 60% TT, 30% TC and CT and 10% CC.

Exercise 5:

The joining of non-homologous ends known as NHEI is responsible for linking DNA sequences that are not homologous to each other. Junctions formed by the NHEJ are commonly found at sites of DNA rearrangements, including translocations, inversions, duplications and deletions. Studies of a large number of these junctions have revealed the presence of 'micro-homology' at junctions ranging from 0 to 0.5 nucleotides. Examples of junctions with 0 and 2 nucleotides of micro-homology are shown in Figure 40.

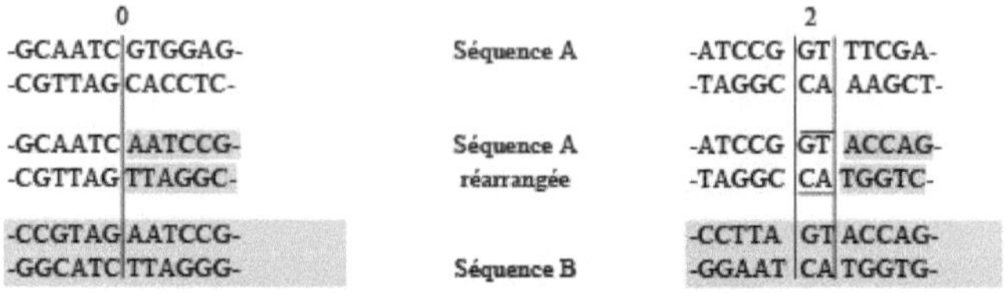

0		2
-GCAATC GTGGAG- -CGTTAG CACCTC-	Séquence A	-ATCCG GT TTCGA- -TAGGC CA AAGCT-
-GCAATC AATCCG- -CGTTAG TTAGGC-	Séquence A réarrangée	-ATCCG GT ACCAG- -TAGGC CA TGGTC-
-CCGTAG AATCCG- -GGCATC TTAGGG-	Séquence B	-CCTTA GT ACCAG- -GGAAT CA TGGTG-

Figure 40: Micro-homology at rearrangement junctions

> Are these short homologies part of the NHEJ mechanism or are they random?

Micro-homologies at 110 NHEJ junctions are shown with their expected random distribution in Table 8.

> Using the chi-square test, determine whether the distribution observed in Table 8 is different from or the same as that expected at random.

Table 8: Observed and expected distribution of micro-homology junctions

Distribution	Micro-homology					
	1	2	3	4	5	6
Observee	47	21	14	13	0	0
Expected	62	31	52	04	0	0

Exercise 6:

Several genes in *E. coli*, including uvrA, uvrB, uvrC and recA, are involved in the repair of UV lesions. *E coli* strains deficient in any of these genes are much more sensitive to the lethal effects of UV light than non-mutant (wild-type) cells for the uvrA and recA strains.

> Why do combinations combining a *rec* A mutation and a *uv* r mutation result in bacterial strains that are extremely sensitive to UV, whereas combinations combining mutations in different uvr genes do not result in greater sensitivity to UV than individual mutations?

For the *uvrArecA* double mutant, a dose of 0.04 Joules/m^2 resulted in 37% survival.

> Calculate how many pyrimidine dimers constitute an intestinal attack for the *uvrArecA* strain. The *E.coli* genome has 4.5 . 10^6 base pairs (assume 50p.100 of GC) knowing that exposure of DNA to UV light at 400 joules/m^2 converts 1p.100 of all pyrimidine pairs (TT,TC,CT, and CC) into pyrimidine dimers.

Solutions to some exercises

Exercise 1:

-A CRM (cross-reacting material or matëriel a réaction croisëe) involves the use of a substance sufficiently different from the reference substance (R) having a function substantially different from the latter, but sufficiently similar to react with anti-(R) antibodies.

1-The presence of CRM suggests that the mutant phenotype is caused by a missense mutation (i.e. the protein is made and can be detected by antibodies but is inactive).

2- Frame-shift mutations result in premature termination of the *putP* and the truncated polypeptide is rapidly degraded by cellular proteases (hence no CRM). On the other hand, a small percentage of spontaneous mutations are base substitution mutations resulting in missense mutations (hence CRM +).

Exercise 2:

-The variable in these experiments is the light: the brighter the light, the less lethality is observed. Visible light can therefore reverse the effects of UV irradiation.

The direct reversal of lesions caused by UV rays is classic in micro-organisms and is known as enzymatic photoreactivation.

The enzyme in *E. coli* has two chromophores that cooperate in capturing photons of sunlight and using their energy to dissociate pyrimidine dimers.

Exercise 3:

A-The adaptive response to low levels of MNNG must require the synthesis of new proteins since the response is blocked by chloramphenicol (see figure 2). If activation of a pre-existing protein were sufficient, chloramphenicol would not be expected to prevent adaptation.

B. The adaptive response may be short-lived for several reasons. Once the signal for adaptation (ultimately, the MNNG) has disappeared, the induced synthesis of new proteins must probably stop.

Resistance to the mutagenic and lethal effects of MNNG then depends on the stability of the induced proteins. If they are relatively unstable, the state of resistance declines rapidly as the proteins become inactive; and even if the proteins are stable the state of resistance of the bacterial population declines rapidly because of their growth and the consëquential dilution of the proteins.

Exercise 4:

A-The extreme UV sensitivity of *UvrARecA* double mutants, compared to cells with mutations at both Uvr gënes, suggestsë that there are two dis- tinate pathways for treating UV damage.

The products of the *Uvr* gene are involved in one pathway, while RecA is involved in a different pathway.

As a general rule, if a combination of mutant genes produces a phenotype that is no more defective than those of the individual mutant genes, the gene products are likely to act in the same pathway.

8. A lethal dose in the *UvrARecA* strain corresponds to approximately one pyrimidine dimer. The number of pyrimidine dimers per lethal dose can be calculated as follows:

Since *E. coli* is made up of 50% GC, the four bases are also represented in the genome. If they had random bacon (which they don't, but this hypothesis is a reasonable approximation), then of the 16 possible dinucleotide pairs in the DNA, a quarter would be pyrimidine pairs. Consequently, the *E. coli* genome (4.6 x 106 base pairs) contains 1.2 x 106 possible UV targets.

Given that a dose of 400 J/m^2 converts 1% of pyrimidine pairs (pyr) into pyrimidine dimers, the number of dimers per dose in E. coli is calculated as follows 100 of pyrimidine pairs (pyr) into pyrimidine dimers, the number of dimers per dose in *E. coli* is calculated as follows:

$$\frac{dimères\ pyr}{dose\ létale} = \frac{1{,}2 \times 10^6\ paires\ pyr}{E.coli} x \frac{0{,}04J/m^2}{dose\ létale} x \frac{1\ dimère\ pyr}{100\ paires\ pyr} x \frac{1}{400J/m^2} 1,2.$$

1-Course reminder :

Transposable ëlëments are discrete DNA segments_that can be inserted repeatedly into multiple sites within the gënome. This process is indëpendent on the previously recognised mëcëcanisms for DNA molecule integration and occurs without the presence of DNA sëquence homology.

Transposons have rëyëкз etre ëlë elements of various size, structure, spëcificitċ of insertion, mëcanism of transposition and regulation and could possëder several phylogënëtic origins.

Simple insertions, on the other hand, are characterised by specific structural features:

- Each insertion contains exactly the same set of non-per- mutëe transpositions of sëquences, they are accompanied by a duplication of a short зë- quence of target DNA;

- The ends of transposons terminate in short inverted rëpëtitions Kleckner (1981) dividedë the transposable ëlëments into three distinct classes, basedë on structural propriëtës, the mëchanism of transposition and the homology of the DNA sëquences of which we know :
- Class I (insertion sequence (IS) modules and the composite ëlëments formedëfrom them. S modules are short ëlëments, smaller than 2 kb in size, encoding only the dsterminants relevant to their own transposition (IS1 {IS5, IS102, ISR1). Two of the copies of certain IS flanking a DNA segment were called composite transposons.
- Class II (the transposon (Tn) family, over 5 kb in size, containing 38 to 40 bp inverted and repeats at their ends, generating repeats of 5 bp of target DNA during insertion.
- Class III contains transposing bacteriophages, such as Mu or its derivatives. These possess genes and transposition sites, as well as genes for DNA rereplication, phage development and cell lysis.

The single insertion sequences (IS) comprise the factors required for cis recombination, in particular, the recombinantly active DNA sequences that define the ends of the segment together with the transposase enzyme that recognises and processes these ends.

Transposable ëlëments have provenërës to be excellent tools for тоlёcи- laire gënëtics:

J As insertional mutagënes, resulting in stable loss-of-function mutations that are easy to map and manipulate.

- *S* As sources of sëlectable markers facilitating mapping, mu- tagenesis, in vitro cloning and construction of recombinant strains.
- *S* As agents for дёпёгег mergers, deletions and inversions of replicons.
- *S* As portable restriction sites for in vitro DNA manipulation, gënëtic mapping and large-scale gënomic mapping.
- *S* As portable homology regions for gënërer fusions, dëlëtions, duplications and inversions of replicons by homologous recombination.
- *S* As vectors to move virtually any gene or sëquence to fixed or al'atory sites in the gënome.
- *S* As ГёШ gene reporters to analyse the control of gene expression, dëtect genes that are only expressed under certain circumstances, the localisation of protëines and their transmembrane domains
- *S* As transcription terminator carriers for analyses of opëron organization and transcriptional regulation.
- *S* As carriers of regul's promoters to gënërer conditional mutations;
- *S* As mobile primer binding sites for sëquen9age of clon' **2-Pre-requisite** DNA**:**
- *S* Have an idea of transposons, their structure and role.
- *S* Know the concept of g'n'tic cartography.

J Know the different types of transposons.

3-Objectives :

- *S* Know the applications and uses of transposons.
- *S* Predicting crosses and recombination results.

Learn how to map.

TD series n :5 "Transposons and gene mapping".

Exercicel :

Transposon Tn10 encodes resistance to tëtracycline (Tet). It is a complete transposon that encodes its own transposase.

> Dëdescribe two ways of transmitting the Tn10 transposon to recipient cells in order to sëlect ëvë transposition events.

Exercise 2:

Opëron srl is required for growth in the presence of sorbitol as the sole carbon source.

Given that a lysate of phage P1 dëveloppë on a random group of aleatory insertions of Tn10 in the chromosome of wild-type *Shigella sp* and a Srl recipient,

> How can you isolate a Tn10 insertion linked to operon srl?

> Describe the crosses you would make and indicate the phenotype of each possible class of recombinant obtained.

Exercise 3:

A common method of transmitting transposons is to use a 'suicide plasmid' which cannot replicate in recipient cells. The pGP704 is an AmpR plasmid that can only replicate if the Pi protein (encoded by the pir gene) is present in the recipient cell.

Given an *E. coli* pir strain^{+} carrying a derivative of pGP704 with transposon Tn5 (Kan),^{R}

> How can you select insertions from an *S typhi* recipient who is r-m +? (Describe the recipient you would use, how to transfer the plasmid and how to select Tn5 insertions).

> How can you easily deduce that the transfer of Tn5 leads to transposition into a large number of different genes on the *S. typhi* chromosome? Explain your logic.

Exercise 4:

The sefA gene encodes a type of pilus unique to *Salmonella enteritidis*. Given a sefA strain^{+}, a sefA ::Kan mutant and a P22 generalized transducer lysate developed on a random pool of Tn10 transposon insertions (tetracycline resistant)

> (Draw a diagram showing the donor and recipient DNA with any ëvë transposition or recombination event, and indicate the medium you would use for each sëlection.)

With a P22 phage lysate grown on a strain with a Tn10 Hëc insertion at the sefA gene^{+} ;

> How could you isolate point mutations in the sefA gene? (Draw a figure showing the donor and recipient DNA and describe the selections you would use).

Exercise 5:

Transposons carrying antibiotic resistance genes can be used as convenient genetic markers: it is possible to require transmission of the transposon by screening for antibiotic resistance, and it is possible to test for loss of the transposon by screening for antibiotic sensitivity. For example, transposons can be used for gene mapping, as illustrated in Table 9.

P22 lysates grown on two put operon-linked transposon insertion mutants (zcc :: Tn10) were used to transduce putP and putA mutants as indicated in the table below:

Table 9: Number of transductants in culture of P22 lysates with two put operon-linked transposon insertion mutants (zcc :: Tn10)

Donor	Receiver	Selection	Transducers			
			Tetr	putP+	Kan^{s}	Put+
Zcc-4 ::Tn10 *put*$^{+}$	*putP* putA/Mud *J*(Kan)^{r}	Tet^{r}	200	35	40	35
Zcc-4 ::*Tn10 putP*$^{-}$	*putP*$^{+}$ putA/Mud *J*(Kan)^{r}	Tet^{r}	200	165	40	40
Zcc-6 ::Tn10 *put*$^{+}$	*putP* putA/Mud *J*(Kan)^{r}	Tet^{r}	200	100	80	80
Zcc-4 :: *Tn10 putP*$^{-}$	*putP*$^{+}$ putA/Mud *J*(Kan)^{r}	and^{r}	200	100	20	0

> Draw a map showing the frequencies and relative locations of cotransduction of zcc-4 :: Tn10, zcc-6 :: Tn10, putP and putA.

Solutions to some exercises

Exercise 1:

The Tn10 transposon can be transmitted using the following two approaches:

(l)Introduce the transposon into a receptor cell using a dëfected phage that cannot replicate in the host nor lysogënëise the host by sëlecting resistance to Tëlra- cyclin.

(2) Introduce the transposon into a receptor cell on a suicide plasmid that cannot replicate in the host by sëlecting Tet resistance.

Exercise 3:

1-In figure 41, a description of the strategy of sëlection of insertions in *S typhi* as a receptor strain is prësentëe;

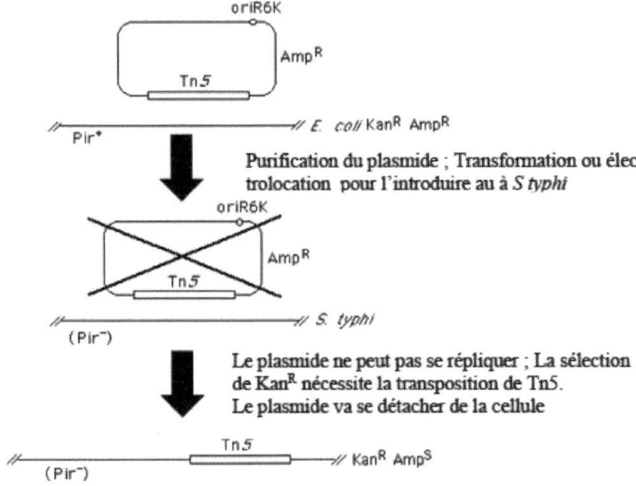

Purification of the plasmid; Transformation or elec- trolocation to introduce it into *S typhi*
The plasmid cannot replicate; selection of KanR requires transposition of Tn5.
The plasmid will detach from the cell

Figure 41: Proposed plan for sëlection of insertions in *S typhi*

B- Explanation :The trick is to look for insertions in a variëtë of gënes that represent a large target site (so it's easy to find insertions in the site) and fa- ciles to dëtect.

Screening for auxotrophic mutations is an excellent way of doing this; auxotrophic mutants represent around 1 to 4% of the 'non-essential' genes in enteric bacteria, and it is easy to screen for auxotrophs by replicating them on a minimal medium.

Exercise 4:

A-Proposal of a diagram describing how to isolate a strain with a
Tn10 near the sefA mutation :: Kan

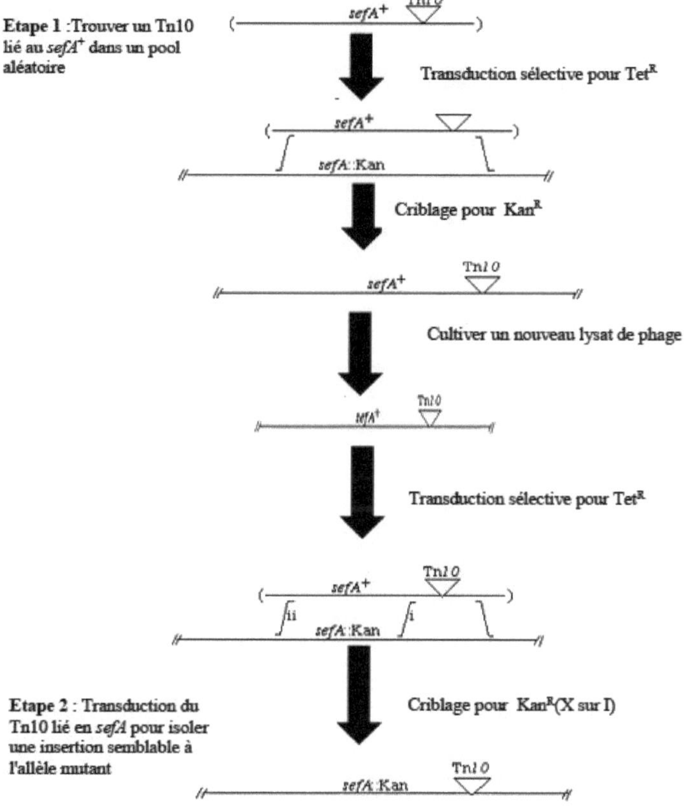

Step 1: Find a *sefA+-linked* Tn10 in a random pool
Screening for KanR
Growing a new phage lysate
Selective transduction for TetR
Selective transduction for TetR
Step 2: Transduction Screening for *sefA-linked* KanR(X on I) Tn10 to isolate a mutant allele-like insertion

Figure 42: Diagram showing the isolation of a strain with a Tn10 insertion and the sefA mutation :: Kan ion similar to the mutant allele

---

b-Diagram showing how to isolate point mutations in the sefA gene:

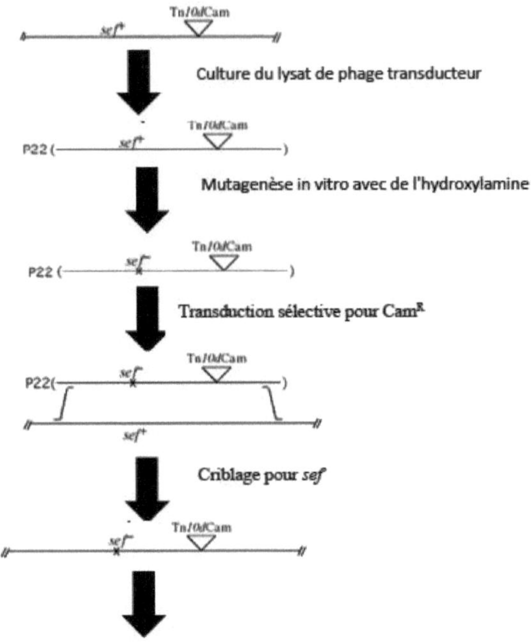

Culture of transducer phage lysate
In vitro mutagenesis with hydroxylamine
Selective transduction for CamR
Screening for *sef-*

Finally, the transducing phage must be cultured and back-crossed to confirm that the mutation is in *sef*

Figure 43: Isolation diagram for a strain with point mutations in the *sefA* gene

1-Course reminder

This phënomëne occurs in a large number of bacterial species (Enteroba cteries, Rhizobium,*Streptococcus, Streptomyces...).* It has been particularly well studied in *Escherichia coli.*

J Conjugative plasmids :

They involve contact between two cells, with a specific pairing taking place between a donor cell and a recipient cell. The donor cell has parietal structures involved in the pairing and transfer of genetic material, known as sexual pili.

These pili belong to 2 categories, F and I, which can be distinguished by lysotypy (specific RNA bacteriophages) and immunology. After contact, pili are liable to retract, causing cells to stick together. The formation of pili is linked to the presence of conjugative plasmids in the donor cell.

In principle, the receptor cells must be of the "female" type, i.e. they must not themselves possess conjugative plasmids. (Remember that there are incompatibilities between the different classes of plasmids).

Plasmid transfer can result in the recipient cell acquiring new properties from the transmitted genes, such as resistance to an antibiotic (this poses a real problem in the case of pathogenic recipient strains).
It should be noted that plasmids carrying resistance (factor R) are classified into 2 categories, which use pili I or pili F for their transfer: pili I are induced by the colicinogen factor ColEl (*pColEI*), while pili F are induced by factor F: both types bind various phages.
Transfer of a conjugative plasmid DNA involves replication by a rolling circle mechanism.

F factor :

Escherichia coli factor F, or fertility factor, is an episome-type conjugative plasmid, i.e._capable of integrating with the chromosome.
Factor F has genes for autonomous replication, for the formation of sex pili, for conjugative transfer and contains insertion sequences (2 IS3 type and 1 IS2 type).
The cells which possess the F factor are called F+: these are the donor cells or "male" cells. The number of F factors per cell is low: 1 to 2.
Receptor cells, or "female" cells, are known as F- cells. During cell contact via the pili, the F cell$^-$ receives factor F and is therefore transformed into an F .$^+$

J Transfer! chromosomal :

During a conjugative phënomëne. and in parallel a Гёоbапдё of F plasmids, which takes place for a large number of cells, a much rarer ëchange of chromosomal gëriel (low-frequency chromosomal conjugation) can occur (frequency 10).$^{-4}$
The F+ cell acts as a donor and the female cell as a recipient: the donor transmits only part of its genetic material (exogënote). Gënëtic mixing is only effective after a process of gënëtic recombination between this exogënote and the endogënote of the recipient cell.
There are mutants that give a high recombination frequency. These are called Hfr mutants: they are derived from F+ cells. In Hfr mutants, the F factor is integrated into the chromosome by a process similar to that of bacteriophage ^ integration.

2-Pre-requisites :

The student must have knowledge of :
The notion of recombination and conjugation.
The conjugation process and its elements.
The concept of conjugative and recombinant plasmids .

3-Objectives :

Learn how to draw up an isolation plan.
Locate the position of the mutations.
Draw a diagram illustrating the recombination event to support your logic.

TD n :6 " Conjugation " series

Exercise 1:

The *proA, proB* and *proC* genes are required for proline biosynthesis. Given a donor strain with an integrated Hfr between the *proA+ proB+ and* proC+ genes as shown below and a *proC* receptor$^-$ Str^R (Figure 44).
> How would you insulate an *F* proC+? (Schematise the proposed insulation plan)

proA· proB· *proC*$^+$

Figure 44: Location of genes genes *proA· proB· and proC·*

Exercise 2:

Str^R is a new mutant that has been isolated and is unable to use acetate as a carbon source (ace). To determine where the mutation occurred, it was crossed to the four different Str^S *ace*$^+$ Hfr donor strains shown below (Figure 45).

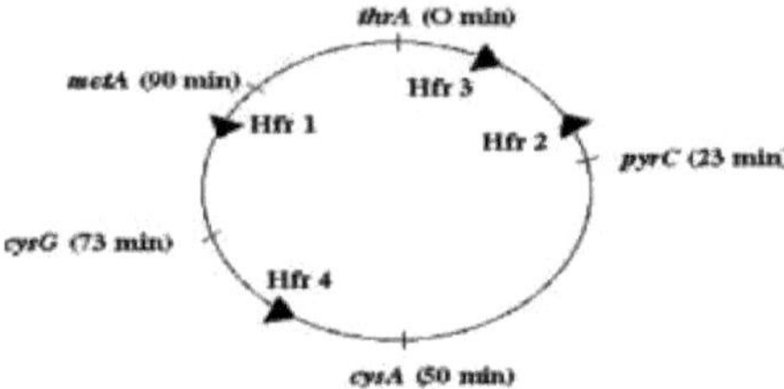

Figure 45: Location and direction of transfer of each Hfr from Str^s *ace*$^+$

> a-What is the selection of exconjugants in this experiment?

> b-What is the counter-selection against donor strains in this experiment?

> c-Analysing the results in Table 10, determine the position of the Ace mutation.

Table 10: Number of ACE colonies$^+$ obtained against each donor strain

Donor strain	*ACE* Colonies
Hfr1	1000
Hfr2	5
Hfr3	1000
Hfr4	80

Exercise 3:

A cross *was* made between Hfr H *met* AziR TonR Str^S and F- *thr leu lac gal* Azi^S Ton^S StrR on plates containing a minimum amount of glucose containing methionine, azide and streptomycin (Table 11).

> a-What markers have been selected?

Recombinants from the above cross were spread on nutrient broth plates containing the indicator dye and carbon source shown below.

> b-What are the genotypes of the recombinants with respect to these carbon sources?

Table 11: Colony colour of recombinant crosses

Source of carbon	Indicator dye	Colour of recombinant colonies
Lactose	Tetrazolium	White
Galactose	EMB	Red gold Violet

Exercise 4:

A strain of *E. coli* resistant to the antibiotics tetracycline (Tet^R) and kanamycin (Kan^R) but sensitive to nalidixic acid (Nal^S) was isolated.

To determine whether antibiotic resistance was encoded on a conjugal plasmid, a cross of the Tet^R Kan^R donor with two different *E. coli* receptor strains, Tet^S Kan^S NalR, was performed. The results obtained are shown in Table 10, together with the "donor only" and "receptor only" controls.

> By analysing the results in table 12, what can you conclude about the transfer of the TetR and KanR genes?

> How can we explain the difference between the results obtained for the two recipient strains used *(E coli* B and *E coli C)*?

Table 12: Crossing results of Tet^R^ Kan^R^ donors with two *E. coli* receptor strains

Donor strain	Receptor strain	Number of colonies on plates with	
		Tet+Nal	Kan+Nal
TetR and KanRNalS	No	2	5
No	*E coli* B TetS and KanSNalR	0	0
TetR and KanRNalS	*E coli* B TetS and KanSNalR	7	4
No	*E coli* C TetS and KanSNalR	0	0
TetR and KanRNalS	*E coli* C TetS and KanSNalR	500	80

Exercise 5:

The chromosomal organisation of *Salmonella typhimurium* and *Salmonella typhi* is shown in Figure 46. Overall, the DNA sequence of the genes of these two bacteria is approximately 99% identical. The *rrn* operons code for ribosomal RNA. Although the location of the RNA operons differs between these two bacteria, the order of the genes located between the RNA operons is conserved (for example, the order of the genes in fragment A is the same in both bacteria).

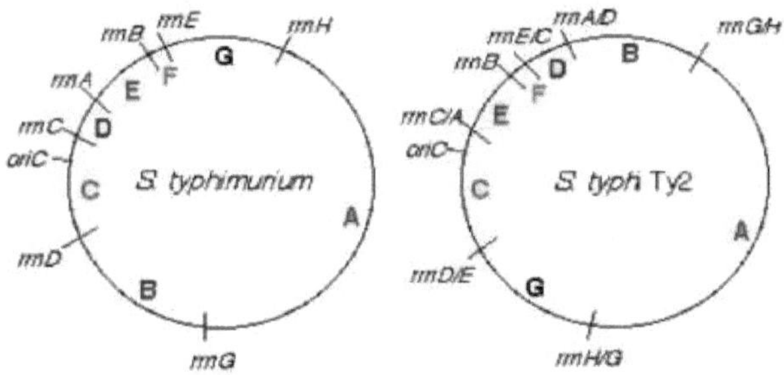

Figure 46: Chromosomal organisation of *S. typhimurium* and *S. typhi*

To enable the isolation of hybrids between these two bacteria, a plasmid was constructed with the following properties: (i) the origin of replication was obtained from plasmid R6K and therefore requires replication of a protein encoded by the *pir* gene$^+$; (ii) the plasmid does not carry the *pir* gene$^+$ but the Pi protein can be provided in trans; (iii) the plasmid contains a clone with the promoter and the first half of an *rrn* operon; (iv) the plasmid encodes Amp^R .Hfr clones from S. *typhimurium* were used to construct them.

> Draw a diagram showing how a possible Hfr would be formed (mention the relevant genes on the plasmid and chromosome and any selection required).

> If the Hfr were integrated into the *rrnB* operon of *S. typhimurium* with an orientation such that it is transferred in a clockwise direction, what is the maximum amount of *S. typhi* DNA that you would be able to replace with *S. typhimurium* DNA after crossing over?

Solutions to some exercises

Exercise 1:

The *F'proC+* will form as shown in figure 46. As the proC gene would be the very last region to be transferred by Hfr, it would only be transferred very rarely. Therefore, we can select for such rare events by associating it with an auxotrophic proC and selecting for growth on a minimal proline-free medium.

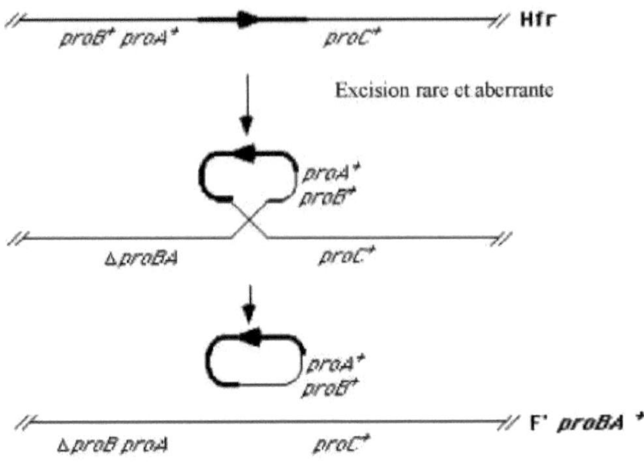

Figure 46: Proposed insulation plan for insulating an *F'* proC⁺

Exercise 2:

A-The selection of exconjugants in this experiment :

-Growth on acetate as the sole source of C.

b- counter-selection against donor strains in this experiment

C-Position of the Ace mutation :

-Approximately 95 minutes, in the total transferred region between Hfr-1 and Hfr-5 (see figure 6).

Exercise 3:

A- Transfer of the TetR and KanR genes:

Since they are transërës at different frequencies, they are probably present on different DNA molecules (for example, separate plasmids).

B- Probable reason for the difference between the results obtained with the two different receptor strains:

This may be due to the restriction barrier in *E. coli*. Remember that *E. coli C* has a lower restriction than *E. coli B*.

1-Course reminder :

The term transduction refers to the transfer of cellular DNA or RNA from one cell to another by infection with a viral vector. Viral particles that carry and introduce cellular nucleic acids during infection are called transducer particles. Bacteriophages are transducing viruses if they carry parts of bac- tërian or plasmid chromosomes from one bacterial host to another.

For some phages, this DNA is linked to the viral chromosome of the particle (as in specialised transduction by phage X.), whereas for other transducing phages, only the DNA of host origin is present in the particle (as in the case of generalised transduction by phages P1 or P22).

The transduction process is divided into two types. In general transduction, the virus can introduce any region of the donor chromosome into a recipient.

In specialised transduction, the virus always transports the same segment to the recipient. A phage that randomly encapsulates host DNA in particles is a generalized transducing phage; a phage that has had a particular gene stably introduced into its chromosome is a specialized

transducing phage.
The ability of a phage to carry out generalized transduction depends on the mechanism by which DNA is encapsidated in phage particles. If this mechanism allows the DNA of the host and the phage to be encapsidated, generalised transduction invariably occurs.
A well-studied example of generalised transduction is the *Salmonella* phage P22. This is a useful paradigm for DNA metabolism leading to generalised transduction, as it is not unique and is representative of a number of other phages (T4) The phage DNA in a P22 particle is terminally redundant and circularly permuted . During P22 infection, the host DNA is not degraded (unlike infections with virulent phages such as T4) and is therefore potentially a substrate for DNA packaging.
For P22, although there are roughly equal amounts of phage and host DNA in the cell during infection, only a few per cent of the host chromosome is packaged and only a few per cent of the particles are transducers. What limits the formation of transducers?
Phage P22 is a useful modële, both for illustrating a common mode of DNA metabolism and for understanding its consequences for transduction. However, among the phages that perform generalized transduction, there are both minor variations on this context as well as quite different modes of transduction (Phage PI ,P22, Phage T4,Phage Mu... etc).
The *E. coli* PI phage is also noteworthy because it is the main generalized transduction phage used in *E. coli* and can infect a wide range of hosts, making it an important tool for genetic manipulation. In generai, PI is similar to P22 by using a delirant conditioning process starting at a specific site to produce terminally permuted and redundant phage DNA.
One of the most important uses of generalized transduction is to measure the distance between markers or to determine the order of three or more markers in the chromosome. An important concept for this analysis is that of counter-transduction, the ability of two markers to be simultaneously integrated into the recipient chromosome on the same transduction fragment.
Several uses for generalized transduction are currently known. These include gene mapping, complementation analysis, plasmid transduction, selection of deletions in plasmids and selection for recombination between clone genes and the bacterial chromosome. Other common uses include strain construction, transposon delivery and isolation of chromosomal rearrangements such as duplications.

2-Pre-requisites :

Understand the concept of bacteriophage.
Have an idea of the term transduction.
Understanding the transduction mechanism.

3-Targets:

Learn about the applications of generalized transduction.
Understanding the concept of co-translators and their usefulness.
Learn how to make a genetic map and determine the order of genes.

Series 7 "Generalized Transduction

Exercise 1:

Plasmid F is approximately 100 kb in size. Since the P22 HT phage contains only about 45 kb of DNA, it cannot cover the entire F plasmid. Therefore, when a tetracycline-susceptible (TetS) recipient is infected with P22 HT grown on a strain carrying an F plasmid labelled with tetracycline resistance (Tet^R), it is possible to obtain transduced strains resistant to tetracycline (Tet).^{R}

> Suggest a likely explanation for this result.

Exercise 2:

To avoid obtaining lysogens resistant to subsequent infection, a virulent mutant of the P1 phage is often used for general transduction. However, this poses a second problem because any cells that are co-infected with both a transducer particle and a phage particle will lyse, preventing isolation of the transducers.

One way of avoiding lysis by the virulent phage is to mix a solution of Ca^{+2} and Mg^{+2} recipient cells and the phage lysate at a low multiplicity of infection (MOI), in-cubate for 15 to 20 minutes, then add citrate to chelate the divalent cations and spread the mixture on a selective medium.

> Why is the mixture left for 15-20 minutes before the citrate is added?

> Why is a divalent ion chelator added to the mixture?

Another way of limiting lysis by the virulent phage is to irradiate the lysate with UV light before mixing the phage with the recipient cells.

> How would this reduce lysis of the recipient cells?

Exercise 3:

Transduction using phage P1 was carried out to determine the order of the *fadD, gap* and *pabB* genes in relation to each other. The donor was *fadD gap*$^{+}$ *pabB* and the recipient was *fadD*$^{+}$ *gap pabB*$^{+}$. Gap transducers^{+} were selected (Table 13).

Co-transmission of non-selected markers is shown in the table below.

> On the basis of these results, what is the predicted order of these genes?

Table 13: Estimated number of colonies

fadD	***pabNumber of***	**colonies**
	++98	
	--14	
+	-04	
-	+19	

Exercise 4:

The *putA900* mutation is a point mutation. The *put-557* mutation is a small deletion in the *put* genes. The *put-544* mutation is a large deletion that removes the entire *put* operon. P22 HT was grown with each of these mutant strains and used to transduce recipients with a *pyrC* or *pyrD* mutation, selecting for *Pyr*$^{+}$. The results obtained are summarised in Table 14.

Table 14: Put- co-transduction results from the two phages P22 and P1

Donor	**ReceiverPhenotype selec-**	**Co-transduction**	**Co-transduction**	
				tionnePut- from thePut- from the
	phage P22phage	**P1**		
	PutTotal	**PutTotal**		
put-900	*pyrC* Pyr+	4500256500		
	pyrD Pyr+	050020500		
put-577	*pyrC*	Pyr+	3	500
	pyrD	Pyr+	0	500
put-544	*pyrC*	Pyr+	100	500
	pyrD	Pyr+	10	500

> Explain why cotransduction between the *pyrD* and *put* genes was observed when the P22 donor contained *put-544* but not in the other two cases?

> Explain the difference in cotransduction frequency observed between the use of phag P1 and P22 as translator phages?

> Will there be a change in the frequency of co-transduction if the recipient strain is affected by the deletion instead of the donor strain?

Exercise 5:

In *E. coli*, the genes for leucine biosynthesis (*leu*) and the ability to ferment arabinose (*ara*) are approximately 50% co-transducible by phage P1. Four different auxotrophs (*leu*) were isolated. Mutations were mapped by P1 transductions. Donor strains were *ara*$^{+}$ *leu* and recipients were *ara leu*. The transducers were selected on a minimal medium containing glucose as a carbon source, then spread on a replicate to test their ability to ferment arabinose (Table 15).

> Which recombinant gënotype did ële sëlectnë initially for these transductions?

> Draw each of the two-factor crosses, showing where crosses must occur between the recipient chromosome and the donor fragment to form Leu transducers^{+} Ara^{+}

Table 15: Proportions of transducers in a two-factor cross

Receiver	Donor			
	ara^{+} *leu-1*	*ara*$^{+}$ *leu-2*	*ara*$^{+}$ *leu-3*	*ara*$^{+}$ *leu-4*
ara leu-1	0%	55%	2%	46%
ara leu-2	3%	0%	3%	2%
ara leu-3	40%	49%	0%	54%
ara leu-4	1%	50%	2%	0%

> On the basis of two-factor crosses, what is the order of the four *leu* allies in relation to each other and the *ara* locus?

Exercise 6:

A sërie of transductions with P22 a ë1ë rëalisëe in order to construct a map for the *met, thi, trp* loci in *S. typhimurium*. The transduction donees are prësentëes in Table 16.

Table 16: Percentage of *met, thi, trp* loci transducers in *S. typhimurium* in the presence of P22

Donor	Receiver	Selected marker	Percentage of Marker	
met^{+} *thi*$^{+}$	*met*$^{-}$ *thi*$^{-}$	Met+	Thi+	80
			Thi^{-}	20
met^{+} *trp*$^{+}$	*met*$^{-}$ *trp*$^{-}$	Met+	Thi+	10
			Thi^{-}	90
met^{+} *his*$^{+}$	*met*$^{-}$ *his*$^{-}$	Met+	Thi+	20
			Thi^{-}	80
his+thi+	*his*$^{-}$ *thi*	Met+	Thi+	10
			Thi^{-}	90
trp+thi+	*trp'*$_{th-+}$	Met+	Thi+	0
			Thi^{-}	100

> Draw a gënëtic map based on the co-transduction frequencies.

> Dëcrivez une transduction suppémentaire qui vous confirmea de plus 1 ordre de carte prevue.

Solutions to some exercises

Exercise 1:

Explanation:

It should be remembered that plasmid transduction requires the encapsidation of a linker concatëmëre of plasmid DNA and its recircularisation in the host cell. Thus, the simplest explanation is that Tet^R colonies arise by transduction of rare and spontaneous F plasmid deletions.

The deletion mutants should reduce the size of the F plasmid to less than 45 Kb (allowing sufficient terminal redundancy for recircularisation by homologous recombination in the recipient), without removing the Tet $marker^R$ or the genes required for vegetative replication.

Exercise 3:

The rarest class of recombinants was Gap^+ *FadD* $PabB^{-+}$. The heritable receptor marker for this strain is $pabB^+$, suggesting that this is the middle marker (Figure 47). Thus, the order of the genes should be: *fadD gap pabB*.

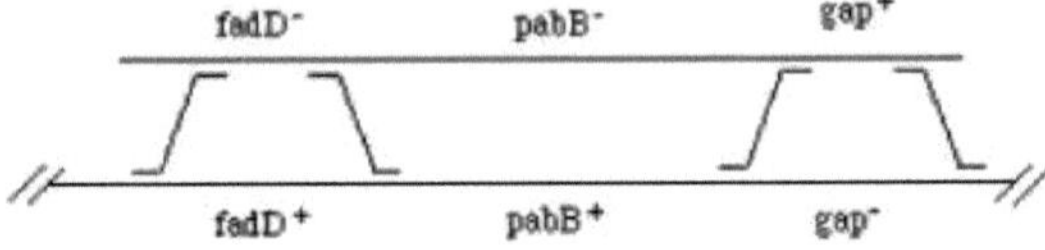

Figure 47: Order of genes

We also know that the three-factor cross data allow us to calculate the percentage of co-heritance between the selected marker and each of the non-selected markers.

When Gap + was selected, the total number of colonies obtained was 180.

PabB's co-heritage⁻ was 54 + 19 = 73 or 73/180 = 41%.

FadD's co-heritage⁻ was 54 + 9 = 63 or 63/180 = 35%.

From these data, the map thus obtained is prësentëe in figure 48:

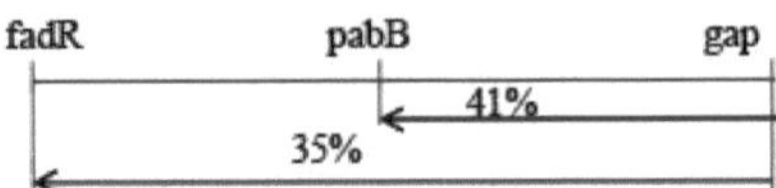

Figure 48: Genetic map of the three recombinants

Exercise 5:

a-Which recombinant genotype was initially selected for these transductions?

-It was the recombinant Leu+.

b- The results of two-factor crosses are shown in Figure 49.

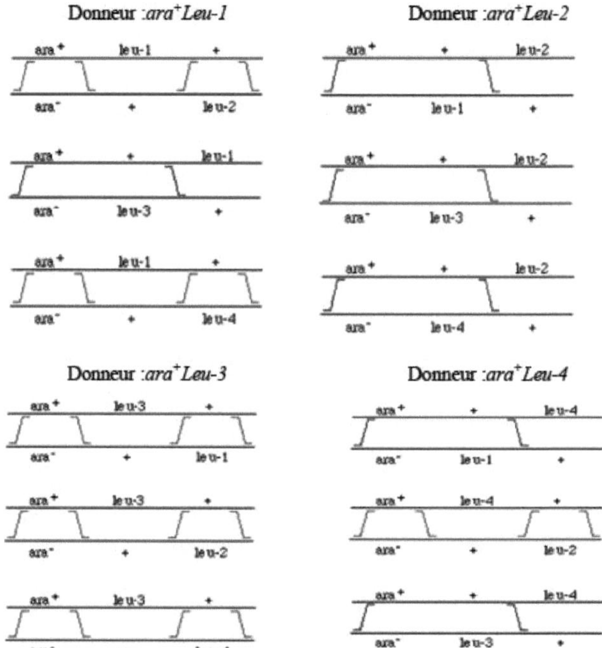

Figure 49:Two-factor crosses of leu transcuctants+ ara+

C-On the basis of two-factor crosses, the order of the four leu alleles in relation to each other and the ara locus will be as follows:

Ordre :*ara ...leu-3... leu-1... leu-4 .... leu-2*

Order: *ara ... leu-3... leu-1... leu-4 .... leu-2*

Exercise 6:

A-The resulting gënëtic map is shown in Figure 50.

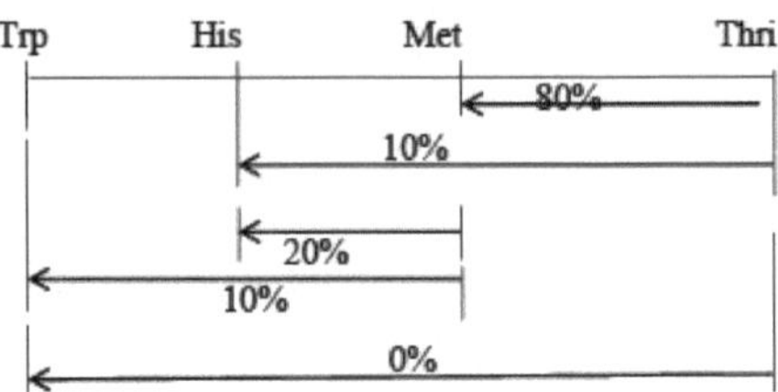

Figure 50: Genetic map

B- The frequency of co-transduction between trp and *his* also confirms the data from two-factor crosses. These data could be obtained by growing the phage on a *Trp* donor$^+$ *his*$^+$, subsequently transduced to a *trp* recipient$^-$ *his*$^-$, by selecting Trp$^+$ or His$^+$. Alternatively, crossing three factors will give more definitive and precise results.

1-Course reminder :

Spëcialisëe transduction is the second major class of virus-mediated gënëtic exchange. It differs from generalized transduction in two respects. Firstly, in this mode of transduction, the transduced genes are covalently linked to the viral chromosome, allowing them to be replicated, packaged and introduced into a recipient along with the rest of the viral chromosome. Secondly, a specialised transduction particle carries a specific chromosome segment and systematically introduces only that set of genes into the recipient. Hence the name specialised transduction.

Phage tempere 1 is the classic example of a specialised transducing phage. The phage 1 particle contains a double-stranded linear DNA molecule with complementary single-stranded ends of 12 nucleotides.when phage tempere 1 infects a cell, its chromosome circularises by hybridisation of these single-stranded extensions (cos sites) and the phage then chooses between two alternative life cycles (lytic cycle and lyosogenic cycle).

Specialised transduction phages are one of the most practical ways of manipulating genes. They represent the original method of gene cloning, using in vivo genetic techniques. Since these phages carry a small region of the chromosome, the genes carried by the phages can be mapped with great precision, as can the mutations within these genes. In addition, mutagenesis can target specific genes. These phages also provide a convenient means of constructing diploid strains for a very limited region of the chromosome for complementation analysis.

2-Pre-requisites :

It is advisable to have an idea of :

J The concept of specialised transduction.

J The similarities and differences between the two types of transduction.

-*S* To find out more about phage 1.

3-Objectives :

J Understanding the fields of application of specialised transduction.

J Precise location of mutations and affected sequences.

J Suggest strategies and alternatives.

Series No. 8 "Specialised Transduction

Exercise 1:

There is a secondary binding site for the lambda phage in the *proB* gene of l'opëroп *proBA from E. coli.*

> Draw a diagram showing how a prophage inte'gre a this site could дё^гег spëcialisëes transducer *strainsproA*.

> How can you use this transduction lysate to obtain a high frequency transduction (HFT) lysate carrying *proA+?*

Exercise 2:

Although P22 is a gënëralisëe transduction phage, it is ëalso possible to obtain spëcialisedë transduction dërivës of P22. For example, spëcialisëes transduction particles have ël ë obtained for the *newD* and *supQ* genes if pulled proximo to the P22 binding site on the *S. typhimurium* chromosome (*ataA*).

> How could you distinguish a gënëralisë P22 transducer lysate from a spëcialisë P22 transducer lysate using simple gënëtic tests? (Be sure to give a gënëtic test that identifies each lysate.

NB:

You can use *S. typhimurium* strains with all the gënëtic markers of your choice, bearing in mind that examination of the phage under an electron microscope, DNA sequencing or the use of antibodies are not satisfactory answers) **Exercise 3:**

Four diffërent lambda *dgal* phages were isotized.These phages were then ël ëcrossed against four lambda mutants, sëlecting wild-type recombinants. The lambda *dgal* phages were likewise ël ë crossed against four different *E. coli gal* mutants, sëlecting Gal+ recombinants. The results of these expëriences are prësentës in Table 17.

> Draw the gal-bio zone of an *E. coli* lambda lysogen, specifying the opëron bio, the opëron gal and the lambda prophage.

> On the basis of the results prësentës in the table above, indicate the relative positions for each of the four *gal* mutations.

Table 17: Results of crossing lambda *dgal* phages with different *E. coli gal* mutants

Phage	**Mutant X points**				***Gal* mutants *of E.coli***	
A dgal	*ff*	*X*	*Y*	*Z*	*gal-116 gal-117 gal-118*	*gal-119*
A dgal-11	+	+	+	+	+-+	+
A dgal-12	+	-	-	+	+-+	+
A dgal-13	-	-	-	+	--+	-
A dgal-14	+	+		+	--+	+

+ indicates that wild-type recombinants have ële obtained

- indicates that no wild-type recombinant has ëlë obtained.

> Indicate the relative positions of each of the four lambda mutations on the lambda gënome and show the missing region in each of the lambda *dgal* phages.

Exercise 4:

A common trick used to perform complementation analysis without excess copies of the gene of interest is to clone the gene onto a lysogenic phage and integrate the phage into the chromosome. For example, the *ompC* gene from *E. coli* has been cloned onto a lambda phage and integrated into the *attB* site of the chromosome. In order to investigate complementation between two *ompC* mutations, a strain is constructed with a mutant allele of the *ompC* gene at its normal chromosomal location and a mutant allele of the *ompC* gene at the phage X binding site.

You have two strains *of E. coli*. One strain contains an insertion mutation in the *ompC* gene which results in resistance to the antibiotic kanamycin (ompC ::Kan). The other strain is a merodiploid with a chromosomal null mutation in the *ompC* gene and a lambda lysogen with a wild-type copy of the *ompC* gene.

> How can the *ompC:: Kan* mutation be easily placed in the m'rodiploid strain?

> How can you tell which copy of the *ompC* gene has acquired the insertion mutation?

> If the *ompC :: Kan* mutation was localised to the lambda lysogen, how could you move the prophage to a new strain?

Solutions to some exercises

Exercise 2:

The difference between generalized transduction and spëcialized transduction lies in the ability to transduce DNA from anywhere on the chromosome.

Thus, we can simply test the transduction of several different gënes located at different sites on the chromosome. A simple way of doing this would be to test the transduction of several different auxotrophic mutations in the recipient, by selecting the prototrophy.

Exercise 3:

Below is a drawing of the gal-bio zone of an *E. coli* lambda lysogen, specifying the bio operon, the gal operon and the lambda prophage and indicating the relative positions of each of the four *gal* mutations:

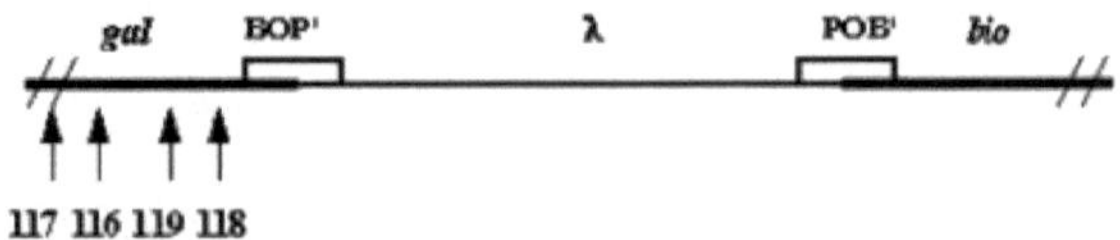

Figure 51: The gal-Bio zone and mutation positions

Indication of the relative positions of each of the four lambda mutations on the lambda genome while showing the missing region in each of the lambda *dgal* phages is summarised in Figure 52.

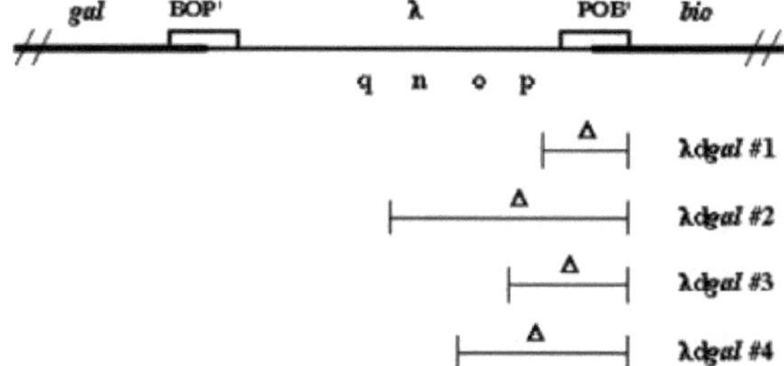

Figure 52: Indication of the relative positions of each of the four lambda mutations on the lambda genome

Exercise 4:

A-To easily dëplacer the *ompC :: Kan* mutation in the mërodiploid strain,grow phage P1 on the *E. coli ompC :: Kan* mutant to obtain a generalized transducing lysate, then transduce the mërodiploid strain sëlectionnëe for KanR.

B- To determine which copy of the *ompC* gene acquired the insertional mutation; If excision of the lambda phage resulted in transducing particles carrying Kan^R , then the mutation must be located in the lambda copy. Alternatively, if the lambda transducing particles could repair the ompC :: Kan insertion in the parent strain, the mutation would not be in the lambda copy.

C- Means of transport :

Reduce the lambda by treating the cells with UV light, then infect the cells with the lambda lysate, selecting Kan .^{R}

1-Course reminder :

A combination of gënëtic and biochemical expëriences on bacteria led to the initial recognition of protein-binding regulatory sequences associated with genes and of proteins whose binding to the regulatory sequences of a дёпе activates or represses its transcription.

These key components underpin the ability of prokaryotic and eukaryotic cells to turn genes on and off, although countless variations on the fundamental process have been discovered.

In the case of bacteria, gene control mainly serves to enable a single cell to adapt to changes in its nutritional environment in order to optimise its growth and division. Research has therefore focused mainly on genes encoding inducible proteins whose production varies according to the nutritional state of the cells.

In microorganisms, the regulation of protein synthesis is highly efficient. Gene expression can be controlled at all levels between the gene and the pro- tein: activation or modification of the gene structure, transcription, maturation of primary mRNA, nucleus-cytoplasm transport in eukaryotic organisms, translation, stabilisation of certain types of mRNA. Most of the control of protein synthesis takes place at the transcription initiation stage: these phenomena have been particularly well studied in prokaryotic organisms.

There are transcription control units known as operons; an operon contains a series of one to ten structural genes, each of which codes for a peptide; these genes follow one another on the chromosome and form a transcription unit that begins with an initiation sequence and ends with a termination sequence.

The initiation sequence is preceded by a promoter, the binding site for RNA polymerase, which controls the transcription of all the genes that depend on it.

The regulation of an operon generally involves an дё^ operator and a дё^ regulator.The operator gene (or site) is adjacent to the promoter. The regulator gene (an independent transcription unit with its own promoter) enables the synthesis of a regulator protein which intervenes at the level of the operator, either by causing the RNA polymerase to bind or pass (negative action) or, on the contrary, by facilitating it (positive action).

There are several types of regulation depending on various factors:

Spëcificitë: control can be spëcific and limited, acting only on the genes of one opëron, or on the contrary non-specific, likely to affect several operons. Non-specific systems are global control systems and generally involve a common regulator.

Nature of the promoter/polvmerase RNA/regulatory protein interaction (positive or negative action).

Final result: this can take two forms. Depending on the case, there is induction in catabolism or repression in anabolism.

2-Pre-requisites :

It is necessary

> *S* To know the concept of the expression of gënes

To have an idëe on the transcription and translation processes.

> *S* To be familiar with the concept of the control unit and the key terms "operator", "operator" and "repressor".

3-Targets

Learn to analyse and determine the cause of a fault and the point at which control is lost.

> *S* Describe, with examples, the impact of mutation on the expression of genes and the correct functioning of catabolic operations.

To understand the utilitë of regulating the expression of gënes.

Series 9 "Regulation of gene expression".

Exercise 1:

> Dëfinish ^gative control and positive control by explaining how the active forms of the protëins that regulate gene expression work.

> Explain how an "inducer" ligand can activate the expression of a gene that is under negative control and how it can activate the expression of a gene that is under positive control.

> Explain how an "inhibitory" ligand can inhibit the expression of a gene under iK'gative control and that of a gene under positive control.

Exercise 2:

In the absence of glucose,*E coli* can proliferate on a pentose sugar,arabinose,using a set of inducible genes organised into three groups on the chromosome.the genes *araA*,araB and araD encode arabinose metabolising enzymes. To understand the regulatory properties of the AraC protëin, you isolate a mutant strain carrying a dëlëtion of the *araC* gene. As shown in Table 17, the mutant strain does not induce expression of the *ara* A gene when arabinose is added to the medium.

A-Based on these results, does protëine AraC regulate arabinose metabolism by iK'gative or positive control?justify your answer.

B-What would the results in Table 18 have looked like if the protëine AraC had regulated the expression of arabinose metabolism enzymes by the type of control you did not choose in question A?

Table 18: Responses of normal and mutant factors to the presence or absence of arabinose

Product of the araA gene

Gënotype	-Arabinosis	+Arabinose
araC+	1	1000
araC-	1	1

Exercise 3:

When Jacob and Wollman tried to vërify the gënëtic link between the gal gene and a prophage of bacteriophage 1 (a gënome л intëgrë), they discovered the surprising phënomene of ërotic induction (which was later called "zygotic induction").

During a bacterial cross, a portion of the chromosome is transferred via a narrow tube from the donor bacterium to the recipient bacterium. Jacob and Wollman discovered that if the transmitted chromosome carried a prophage 1, but the recipient cell did not, the growth of 1 was induced in the recipient cell, hindering its lysis and the production of phage 1.

However, if the recipient cell carried prophage 1, no lysis was observed. A summary of the results obtained from all their crosses is shown in Table 19.

> A-Explain how these results are consistent with the notion that a prophage-encoded repressor normally keeps bacteriophage lytic genes turned off. Justify your answer

> B-Suppose that the prophage prevents lytic growth by expressing a gene expression regulator protein that turns on a gene encoding an anti-lysis protein. Would the results of the crosses have been identical or different? Justify your answer.

Table 19: Results of crosses between bacteria carrying or not prophage 1

	Lambda+ receiver	lambda+
Lambda donor-	No lysis	No lysis
lambda+	Lysis +phage lamba	No lysis

Exercise 4:

A single class of mutations in the crp gene is designated crp*. The properties of the crp* and crp^+ mutations are shown in the table below. Cells were grown in minimal medium with glycerol (as a carbon source) and IPTG (Table 20).

Table 20: в-galactosidase activity of cultivated strains

Strain	**в-galactosidase activity**	
crp^+ cya^+	1000	
crp^+ cya^-		20
crp* cya^+	1000	
crp* cya^-	1000	

> What is the likely molecular mechanism of the phënotype of cap* mutants? Explain your answer briefly.

> How would the crp* mutation affect the expression of other catabolic opërons such as gal or ara?

> How would the crp* mutation affect lac expression in a strain containing a class I lac promoter mutation?

> How about a class II lac promoter mutation?

> If you were to develop a crp* strain in minimal medium + glucose + IPTG, what level of в-galalactosidase would you expect?

> If you were growing a crp* strain in minimal medium + glycërol without IPTG, what level

of в -galactosidase would you expect?

> If you constructed a partial diploi'de (F'crp*/crp$^+$), which mutation would be dominant and why? If you add a column to the table with this mutant, what would be the predicted levels of в -galactosidase?

Exercise 5:

E. coli proliferates faster on glucose (monosaccharide) than on lactose (disaccharide) for two reasons: (1) lactose is prëlevë slower than glucose and (2) lactose must be hydrolysedë into glucose and galactose (by a *в-galactosidase*) before it can be mëtabolisë further.

When *E. coli* is grown on a medium containing a mixture of glucose and lactose, it proliferates with complex kinetics (Figure 53).

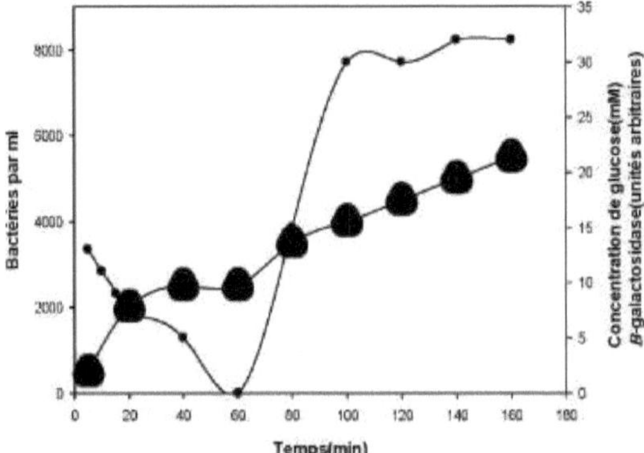

Figure 53: Proliferation of *E coli* on a mixture of glucose and lactose

Bacteria proliferate faster at the beginning than at the end, and there is a dëcalage between these two phases, corresponding to a period during which the bacteria practically stop dividing. When we measure the concentrations of the two sugars in the medium we see that that of glucose becomes very low after a few cell doublings (Figure 17) while that of lactose remains high almost until the end of the experiment. Although the concentration of lactose is high throughout most of the experiment, в- galactosidase is not induced until more than 100 minutes have elapsed.

> Explain the kinetics of bacterial proliferation during the experiment and justify the initial rapid rate, the final slow rate and the delay occurring in the middle of the experiment.

Exercise 6:

Two different mutants with mutations in the attenuator region of the tryphthophan operon were analysed. The first mutation corresponds to a base substitution mutant, while mutant No. 2 is a deletion that deletes certain specific sequences.

The effects of these mutations on the expression of the *trpE* structural gene are shown in Table 21.

Table 21: Effects of these mutations on the expression of the structural gene *trpE*

Strain	*trpE* activity
trpR trpL+ (parent)	1.0
trpR trpL #1	0.1
trpR trpL #2	0.1
trpR trpL #1 Pseudo-mutant	2.5 - 4.0
trpR trpL #2 Pseudo-mutant	2.5 - 4.0

> How to explain the effects of mutations No. 1 and No. 2 at the molecular level

The expression of *trpE* in pseudomutants is also shown in Table 18. Some revertants, in the case of mutant n°1, were true revertants whose wild-type base was restored.

> Assuming that the pseudo mutants are base substitution mutations, where are they likely to map into the RNA sequence shown below?

> How do they affect trp expression at the molecular level?

Solutions to some exercises

Exercise 1:

A-If the active DNA-binding form of a gene expression regulatory protein (a дёпе repressor protein) functions to turn off the expression of gënes, the mode of gene regulation is called negative control.

If the active DNA-binding form of a gene expression-regulating protëin (a gene activating protëin) has the function of activating the expression of gënes, the mode of gene regulation is called positive control.

B-An inducer ligand can activate the expression of a negatively controlled gene by binding to the repressor protein, causing a decrease in its affinity for DNA, allowing it to unhook. In the absence of the repressor, the дёпс' is expressed. An inducer ligand can activate expression of a positively controlled gene by binding to the activator protein, causing an increase in its affinity for DNA, allowing it to bind and thus activate expression of the gene.

C-An inhibitory ligand can inhibit the expression of a negatively controlled дёпс' by binding to the repressor protein, causing an increase in its affinity for DNA which allows it to bind and thus inhibit the expression of the gëne.

An inhibitory ligand can inhibit the expression of a positively controlled gene by binding to the activating protein, causing a decrease in its affinity for DNA and allowing it to detach, thereby inhibiting gene expression.

Exercise 2:

A-The results shown in Table 18 are those expected for a positive regulator of the arabinose operon. A positive transcriptional regulator must bind to the promoter regions of arabinose gënes in order to stimulate their transcription. Thus, if the gëne of the regulatory protein was deleted, the arabinose genes could not be activated.

B. If the AraC protein were a negative transcriptional regulator of gene expression, the arabinose genes would have been fully induced, regardless of the presence or absence of arabinose. Since a negative regulator blocks transcription, its elimination by deletion would have allowed arabinose genes to be transcribed under all conditions. Bacterial operons that are negatively regulated by a repressor (such as the Lac operon) show exactly this behaviour: they are expressed at high levels in the absence of the repressor protein.

Exercise 3:

A-The results of the crosses prësentës in table 16 show that bacteriophage proliferation and cell lysis occur only when the donor carries prophage 1 and the recipient lacks it.

These results are consistent with the idea that a repressor keeps the prophage's lytic genes turned off. In bacteria carrying a prophage, the presence of the repressor ensures that the prophage remains quiescent. When the prophage is transferred to a prophage-free bacterium, it finds itself in a repressor-free environment and its lytic programme is induced.

If, on the other hand, the prophage is transferred to a bacterium which itself possesses a prophage, the presence of the repressor prevents the prophage from being induced.

B-We would expect exactly the same results if lysis were controlled by a gene expression regulatory protein activating the expression of an anti-lysis protein. If the prophage enters a prophage-depleted recipient, the activator is absent, preventing expression of the anti-lysis protein and allowing induction of the prophage and lysis.

It was only with the help of additional genetic and biochemical experiments that the anti-lysis factor was identified as a repressor.

Exercise 5:

A-The rapid bacterial growth at the beginning of the experiment is the result of glucose metabolism. The slower growth at the end is the result of lactose metabolism. The bacteria stop growing in the middle of the experiment because they have run out of glucose but do not yet possess the enzymes required for lactose metabolism. Before they can use the lactose present in the medium, they need to induce the lac operon. The delay observed in growth represents the time required for this induction.

B. Induction of the lac operon requires two conditions to be met: lactose must be present and there must be no glucose. During the first part of the experiment, both glucose and lactose are present, so the conditions for induction are not met. They are only met when the glucose is used up.

CAP and the lactose operon repressor promote induction For the operon to be active, CAP must be bound and the lactose operon repressor must not be. The presence of lactose in the medium increases the intracellular concentration of allolactose, which binds to the lactose operon repressor, lowering its affinity for its binding site and causing it to detach from the DNA. Removal of the repressor of l'opëron lactose satisfies 1 of the conditions for induction. The second condition is Hëc a glucose concentration. When glucose concentration falls, the intracellular level of cAMP rises. The cAMP binds to CAP and changes its conformation so that it can bind to its binding site. When CAP is in place (and the lactose operon repressor is absent), RNA polymerase can bind to the promoter and initiate transcription.

List of References

♦♦♦ ***Bibliography***

1-Black ,J.G.,Black ,L.J. Microbiology: Principles and Explorations 9th Edition.Publisher : Wiley. August .2015-960 pages.

2-Brown.J.W. Principles of Microbial Diversity. Publisher: ASM Press. January.2015-416 pages.

3-Chinnici,J.P., Matthes ,D.J. Genetics: Practice Problems and Solutions 1st Edition.Publisher: Pearson. November .1998- 350 pages.

4-Evans,H. Bacteria: Microbiology and Molecular Genetics. Syrawood Publishing House. May. 2016- 213 pages.

5-Griffiths,A.J.F.,Wessler,S.R., Carroll,S.B., Doebley,J.An Introduction to Genetic Analysis,11th Edition, Publisher: Freeman/Worth,January.2015-896 pages.

6-Hunt,T., Wilson,J .The Problems Book: for Molecular Biology of the Cell ,6th Edition. Publisher: W. W. Norton & Company,November.2014-984 pages.

7-Kowles,R. Solving Problems in Genetics 1st Edition. Publisher:Springer.June.2001- 500 pages

8-Maloy,S ., Cronan,J.E .,Freifelder,D.Microbial Genetics 2ed Edition.Publisher : Jones and Bartlett.April.2004-484pages.

9-McGrath,S., Sindere,D.V. Bacteriophage: Genetics and Molecular Biology 1st Edition,publisher. Caister Academic Press. July .2007-344pages.

10-Pommerville,J. Fundamentals of Microbiology 11th Edition.Publisher: Jones & Bartlett Learning.May .2017-944pages.

11-Russell,P. J.,Chase,B.J.Study Guide and Solutions Manual for iGenetics: A Molecular Approach 3rd Edition.Publisher: Pearson.April.2009- 456 pages.

12-Seshasayee,A.S.N.Bacterial Genomics: Genome Organization and Gene Expression Tools 1st Edition.Publisher: Cambridge University Press.March.2015-230 pages.

13-Snyder,L.,Peters,J.E., Henkin,T.M., Champness,W. Molecular Genetics of Bacteria, 4th Edition. Publisher: ASM Press, January .2013-728 pages.

14-Talaro,K.P., Chess,B. Foundations in Microbiology: Basic Principles 10th Edition.Publisher : McGraw-Hill Education.February .2017- 640 pages.

15-Tortora,G.J.,Funke,B.R.,Case.C.L. Microbiology: An Introduction 12th Edition.Publisher : Pearson.April 2016-960pages.

16-Uldis,N.S.,Ronald ,E.Y. Modern Microbial Genetics, 2nd Edition.Publisher: Willey-Lisse, February.2002-672 pages.

♦♦♦ ***Webography***

http://cgsc.biology.yale.edu/
http://www.asmusa.org
http://www.genome.jp/kegg/
https://sciences.sdsu.edu/
https://www.ncbi.nlm.nih.gov/

Printed by Books on Demand GmbH, Norderstedt / Germany